POLYNOMIAL FUNCTIONS FOR DETERMINING THE TREND IN THE MECHANICAL PROPERTIES OF STEEL DEPENDING ON THE TIME OF CORROSION INFLUENCE

ANTONIO SHOPOV

2025

POLYNOMIAL FUNCTIONS FOR DETERMINING THE TREND IN THE MECHANICAL PROPERTIES OF STEEL DEPENDING ON THE TIME OF CORROSION INFLUENCE

Author: Dr. Antonio Shopov.

Translation from Bulgarian

English, Second edition, Sofia 2025

CONTENTS

CHAPTER I. MAIN REVIEW

1. Introduction

The analysis of basic structural elements, using the traditional approach, consists of two main steps. In the first step, the stresses, deformations and displacements of the structure subjected to external loads are calculated and the limit values of these loads are determined. This problem is solved with the using of the fundamental theory from the disciplines: strength of materials, structural mechanics, the theory of resilience, elasticity and plasticity and the general theory of materials, but engineering calculations do not end there. The ultimate goal is to determine whether the structure will be able to perform its intended purpose reliably enough for a given period of time. The second stage of the calculation is to compare the calculated stresses, deformations and displacements with certain normatively permissible values and to compare them with the design operational loads. Although basic, the second stage of the calculation is also very important, as it involves uncertainty that increases over time due to the presence of a variety of factors [65].

The statistical interpretation of the safety factor allows for a more rational and comprehensive method for assessing reliability. In this case, it is necessary to optimize the first stage of the calculation. In order to assess reliability, it is necessary to know the mechanical properties and characteristics of the used construction products under operational conditions. The random change in time of the external load and impact should be taken into account, taking into account the factors and parameters contributing to this change. With this approach, new unknown components of the system appear, forming the conditions for determining the stochastic properties of random variables, behavior and changes in the characteristics of the material based on these already changed external conditions [65].

In solving this type of problem, methods of statistical mechanics and reliability theory are usually used sequentially (with the exception of some optimization approaches). The boundaries roughly range from determining stresses, strains, etc. to identifying hazardous situations and calculating the probability of their occurrence. In most cases, there are no difficulties at the first stage and the solution of the problem is essentially reduced to calculating reliability. The actual reliability calculations are carried out by means of detailed mathematical calculations to determine the characteristics of stresses and strains for a given structure. However, for methodological reasons, it is necessary to define separately the calculations for statistical mechanics of structures and the theory of their reliability [65].

For various reasons, the problem of structural reliability attracted public attention earlier than the problem of statistical mechanics. The theory of reliability as a general technical necessity took shape around 1960, influenced mainly by the development of radio electronics, computer technology and crane construction. However, for the first time, the problems of reliability theory were directly included in structural mechanics after 1920. The first papers on reliability theory were published by M. Mayer [56] and N. F. Khotsialov [61], criticizing the concepts of allowable stresses and safety factors. In contrast to this concept, the idea of applying statistical methods in strength calculations provides an opportunity to find a solution. In these papers, the basic concepts of reliability theory can be found. The first publications on the reliability of structures are controversial and are not accepted by professionals in this field. An important role in the introduction of statistical methods into structural mechanics was played by I. S. Streletsky [60], who published a series of research papers on the topic since 1935. In [60], the concept of statistical structural reliability is systematically presented. This concept is implicitly reflected in the methodology of structural calculations based on limit states, material behavior theory and safety requirements. After 1945, research in this area continued, such as the research of A. R. Rzhnitsyn [59], A. Freudenthal [53], A. Johnson [54], etc. These developments are distinguished by the desire to simplify computational schemes that do not require complex analytical equipment. The schemes allow obtaining a qualitative description of the phenomenon, studying the influence of load variability and strength on reliability, as well as raising questions about optimization and other aspects of the operational state. During the same period, the introduction of statistical methods in mechanical engineering, shipbuilding and other technological fields also began. In mechanical engineering, reliability issues are considered mainly in connection with the problem of the durability of machine components operating under variable stress conditions. The last two decades have been characterized by a rapid increase in the volume and level of scientific research. The theory is based on three closely related ideas. The first idea is based on a clear understanding of the external operating conditions of the structure and the fact that its behavior during its operation is a random process [65].

The second approach is based on the probability that a given parameter of the system is within a certain permissible range and this is the main indicator of reliability. Violation of normal operation is, in principle, a deviation of the structure from this range, which is interpreted as a consequence of the gradual accumulation of residual deformations, wear and the appearance of other defects. When these defects reach a certain size, the normal operational and computational behavior of the structure is violated. Thus, the application of reliability indicators, characteristic of early research - the probability of satisfying certain inequalities connecting random variables, is replaced by a more comprehensive interpretation based on the theory of random functions [16-67].

The third way is by predicting the changes in the mechanical properties of the material over time depending on the operating conditions. After taking these changes into account, a check should be carried out for the physical and deformation characteristics.

In the late 19th century, Poincaré [13] laid the foundations for discrete nonlinear dynamical systems. In 1963, Edward Lorentz refined this type of dynamical systems [8], building on them, and subsequently in 1975, Lee and York defined the theory of nonlinear dynamical systems as we know it today. This in itself represents the largest upgrade of Newtonian mechanics as we know it, taking more than 300 years.

This symbiosis forms modern methodologies for calculating classical mechanics, using all available tools of applied mathematics.

2. Review in statistical mechanics

Statistical mechanics is defined as the use and application of the mathematical theory of probability to the classical theory of mechanics. Statistical mechanics describes the interactions between a large number of particles (often of the order of Avogadro's number), relating the properties of elementary particles to the properties of macroscopic objects and matter observed in everyday life. The modern theoretical mechanics we know finds its basis in statistical physics [16-52]. The main advantage of statistical mechanics over classical mechanics is that it can explain the properties of matter based on the theory of the interactions between its constituent particles. In both theories there is the idea of entropy, but in statistical mechanics it is a function of the number of possible micro and macro states, while in classical mechanics it is a precisely determined and defined quantity. The subject of statistical mechanics is the study of the behavior of mechanical, electrical, radiotechnical, biological and similar systems under random external influences and/or random changes in the properties of the system (material) [16-52]. It is possible to develop a general theory of the behavior of such systems without specifying their properties and characteristics [65]. At the same time, issues such as the formulation and solution of fundamental problems come to the fore. This determines a more general formulation of many of the results obtained in the theory of statistical communications, the theory of automatic control systems, and other areas of applied mathematics. When considering a system (fig.1.2.1) interacting with its environment, we can first assume that both the properties of the system and its interaction with the environment are completely determined [65]. The hypothesis is adopted that external influences are characterized by an element h in the space (if the condition is met that the space is less than 2), and the behavior of the system is characterized by an element i in the space D. They can be numbers, vectors, tensors, functions of one or more variables, etc. The structure and properties of the system are characterized by the operator H and each realization of the external influence h is associated with a realization of its behavior "u" [65].

$$u = H.q \tag{1.2.1}$$

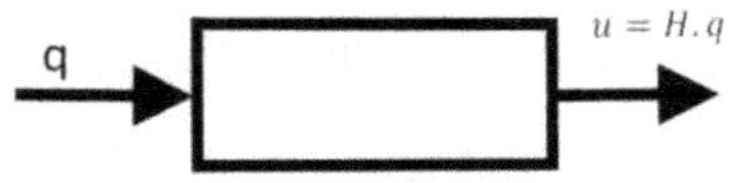

Figure 1.2.1 System schematic diagram

One such system is the elastic system that is subjected to external forces. It is the external forces that play the role of acting forces, and the parameters of the system's behavior - displacements, stresses and strains [65]. The "AND" operator is determined by the equations of structural dynamics and the corresponding initial and boundary conditions [65].

To illustrate the concept and principles, it is easiest to explain it with an example from structural mechanics: We take an elastic straight prismatic rod of length l, loaded with a horizontal force F and a distributed transverse load, the intensity of which is equal to q. These forces are generally functions of time [16-65]. The behavior of the rod is described by the function w(x) coordinates x and time t, i.e. F = F(t), q = q (x, t) и w(x) = w (x, t) – fig.1.2.2.

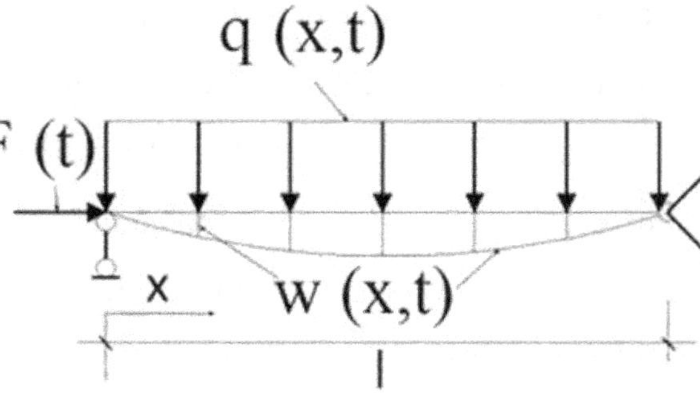

Figure 1.2.2 System diagram

Equal to the transverse displacements are the points that lie on the axis of the rod. Under some additional assumptions, the function w(x, t) should satisfy the equation [65]:

$$EJ\frac{\partial^4 w}{\partial x^4} + F\frac{\partial^2 w}{\partial x^2} + m\frac{\partial^2 w}{\partial l^2} + k\frac{\partial w}{\partial w} = q \tag{1.2.2}$$

where: EJ is the stiffness of the section of the element during bending; m is the mass of the rod per unit length; k is the coefficient of viscous friction. We must consider initial and boundary conditions. These conditions are [65]:

$$\left.\begin{aligned} w = \frac{\partial w}{\partial t} = 0\,,\ \text{in conditions} \rightarrow t = 0, 0 \leq x \leq l \\ w = \frac{\partial^2 w}{\partial x^2} = 0\,, \text{in conditions} \rightarrow 0 \leq t \leq +\infty\,, \text{for } x = 0 \text{ or } l \end{aligned}\right\} \qquad (1.2.3)$$

The differential equation (eq.1.2.2) and the conditions (eq.1.2.3) represent a specific realization of the dependence (eq.1.2.1). In the practical and physical sense, the interaction of the element with the environment can be characterized by the forces F(t) and q(x, t), which play the role of input conditions - depending on the change in the relevant environmental factors. The output process is the displacement function w (x, t). So [65]:

$$w = H[F, q] \qquad (1.2.4)$$

If the longitudinal force F is constant, it must be included in the system. In other words, an elastic beam loaded with a constant force F must be considered as a system on which an external load q(x, 1) acts. In this approach, the force F becomes a parameter of the system and is included in the definition of the special application of the operator H [49-52,65].

Parametric is the influence that can be included in the properties of the system fig.1.2.3.

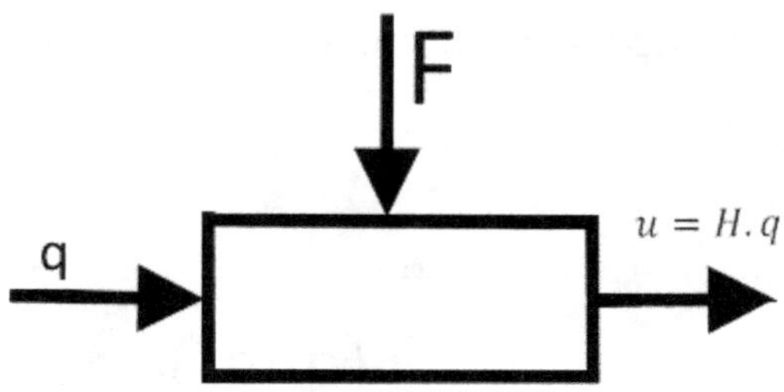

Figure 1.2.3 Process flow diagram

There is another important reason for the uncertainty of the operator dependence in eq. 1.2.1, and it is that the output parameters can be chosen in different ways [48-52,65].

Depending on this choice, the form of the operator H will change. We assume that the space u is exhaustive in the sense that all possible behaviors of the system can be described by its elements [64]. It is well known that for any system, in general, there are an infinite number of exhaustive spaces. All of them are equivalent, since each of them can carry complete information about the behavior of the system

[16-25]. The operators H corresponding to different exhaustive spaces must be expressed from each other by certain relations. An exhaustive space means that for each of its element's **q** it is possible to reconstruct the corresponding element q belonging to the set of numbers [25-36]. Under this hypothesis, there exists an inverse operator L such that it satisfies the condition [65]:

$$L.u = q \tag{1.2.5}$$

It is worth mentioning that the problems of structural mechanics, the theory of elasticity, the theory of vibrations, etc. are usually posed exactly in the form of eq. 1.2.5, i.e. in a form that is not resolved with respect to the elements in a given space U. The approach to inverting the operator L is one of the main tasks in calculating mechanical systems, since in each calculation or study one does not always obtain comprehensive information about the behavior of the system [16-32]. In practical problems it is often sufficient to limit the values of certain parameters to individual points or connections in the system. For this reason, in static calculations of structures they are often limited to determining the stresses only in the most critical elements, and the displacements only in those points where they are expected to be maximum [16-65]. This approach allows for even greater variety in the choice of the given space of numbers (or a numerical sequence) and this has an impact on the operator H. For non-exhaustive spaces, the operator H is assumed not to be expressed in terms of one another, although it can be expressed in terms of one of the operators coinciding in the exhaustive space [65]. Another classic example, to illustrate the convention of separating the system depending on the environment, is the existence of a thin elastic plate loaded with a normal load of intensity q, corresponding to fluctuations in a given gas flow (fig. 1.2.4) [65].

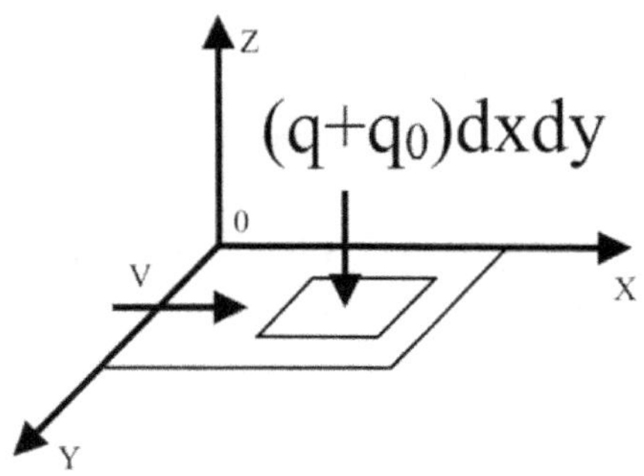

Figure 1.2.4 Process diagram

The vibration equation under certain conditions of this plate is represented by the equation corresponding to the form and it is of the form [65]:

$$D\Delta\Delta w + m.\frac{\partial^2 w}{\partial t^2} + k\frac{\partial w}{\partial t} = q + q_0 \qquad (1.2.6)$$

where: w(x,y,t) is the normal displacement of the points located on the mid-surface of the plate; m is its mass per unit area of the mid-surface; k is the damping coefficient; Δ is the Laplace operator in the (x,y) plane. To solve the equation, we need to know the initial boundary conditions, which can be [65]:

$$\left.\begin{aligned} w = \frac{\partial w}{\partial t} = 0\,,\text{in conditions} \rightarrow t = 0\,,x\,,y\,(-\Omega) \\ w = \frac{\partial w}{\partial n} = 0\,,\text{in conditions} \rightarrow 0 \le t < +\infty\,,x,\ y\,(-\Gamma) \end{aligned}\right\} \qquad (1.2.7)$$

Here, we will assume that Ω is the area occupied by the average surface of the plate; Γ - plate contour ; $\frac{\partial w}{\partial n}$ — derivative of motion along the normal to the contour.

In the interaction of the gas with the plate, we should also take into account the pressure through q_n (x,y,t), which is included in the right-hand side of the equation (equation 1.2.6) [65]. A mandatory prerequisite is that the pressure at each point is a function of w(x,y,t) over the entire surface Ω . The simplified version is to assume that the flow around the plate passes at a high-speed V (let's assume supersonic speed) and the available disturbances (noises) received from it are quite small, and then with sufficient accuracy for practice we assume that [65]:

$$q_0 = q_\infty + \frac{\chi\, q_\infty}{c_\infty}\left(\frac{\partial w}{\partial t} + V\frac{\partial w}{\partial x}\right) \qquad (1.2.8)$$

where: χ is the polytropic index; q_∞ and c_∞ - the pressure and speed of sound in undisturbed flow (respectively).

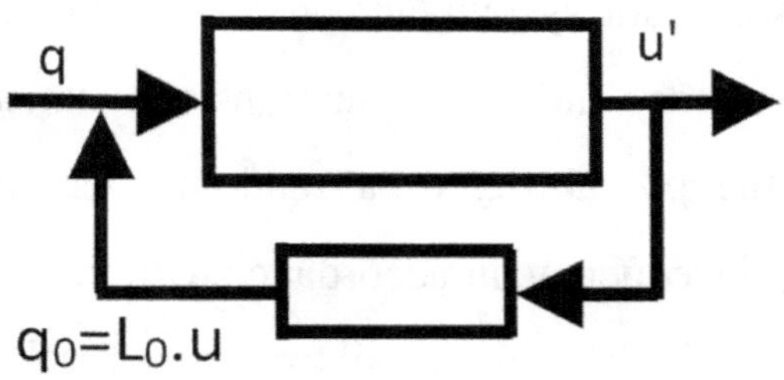

Figure 1.2.5 Process model

Since the gas pressure q0 depends on the displacement, we should treat it not as an input parameter, but as an operator of the process behavior (fig.1.2.5). Thus, we are sure that the unperturbed gas flow is an integral part of the system. We have an operator equation of the form [65]:

$$Lw = q + L_n w \qquad (1.2.9)$$

reduces to the form of eq.1.2.5, if we assume L-L_0 for the system operator.

It is normal to assume or claim that the properties of a given system are deterministic when we can include external influences caused by natural influences in it. Statistical mechanics, as mentioned above, studies the development of a given system under random influences and/or0 under random variables in the properties of the system [16-52]. The above-mentioned methods for the spaces of the input and output parameters and for the operator specificity of the system are completely reliable when considering probabilistic problems. We come to the final premise that in the case of stochastic systems in statistical mechanics, the described parameters change, but the method for describing the system also changes [26-33].

The input parameter q is random, i.e. it represents a random number, a random function, etc. Then each element q belonging to a given space will be associated with some certain probability unit. The input parameter is a random number, then it is characterized by a distribution function (probability density). Also, the input parameter is a random vector, and then it exists as a multidimensional joint distribution for the corresponding components. This is a time function that is random and is defined by a complete system of joint distribution functions of its values at arbitrarily chosen times. Instead of a full definition by indicating the corresponding units in the space of functions, partial names are replaced. This method is widely used in practice with integrals with respect to probability units: mathematical expectations, dispersions and other units of random variables, moment and correlation functions of random processes, etc. The input parameter **q** is random, i.e. it represents a random number, a random function, etc. Then each element **q** belonging to a given space will be associated with some particular probability unit.

The identification of stochastic systems and processes should be considered in more detail in each direction [31]. In probability theory and its applications, the statistical interpretation of probability is widely used. In this case, the probability of a random event is interpreted as an objective unit corresponding to the empirical frequency. In accordance with the law of large numbers, the probability of a random event is the limit to which the empirical frequency (probabilistically) bends when the number of observations increases to infinity [31-32]. When applying the concept of empirical frequency, it is also assumed that random events are large-scale, i.e. that they are generated repeatedly under statistically homogeneous conditions [31-32].

It follows that statistical interpretations of probability can be extended only to systems that apply to a large number of statistically homogeneous, comparable cases. Only after we have a sufficiently representative set of data from relevant systems, statistical estimates of the properties of the entire system are obtained. On the other hand, only under these conditions can a statistical explanation be given to the conclusions reached in solving the problems of statistical mechanics [23-29]. However, this approach will significantly reduce the scope of application of statistical methods. Not only unique (and often the most expensive and responsible) systems will be excluded from consideration, but also systems produced in a relatively small number of comparable trial elements. As already noted, probability is an objective measure of the likelihood of an event occurring, regardless of whether it is comprehensive or not [29].

Analyzing this approach is not a technical, but a philosophical-logical and/or psychological problem. To avoid the difficulties associated with this, one can approach it using the concept of possible coordination between the components of the system under consideration, i.e. a set of hypothetical comparable systems considered together with a given system [23].

Despite the above concept, there will still be considerable practical difficulties. Appropriate statistical information is needed to obtain probabilistic properties of the system, and this can generally only be obtained by analyzing a representative set of studies [23]. These studies must be real, not theoretical. Similar difficulties arise when the system is very large [23].

The corrosion influence is caused by the diverse properties of the various systems, which are ergodic, i.e. they manifest themselves not only in the relevant limits of realizations, but also in time and/or space. Such effects can be studied by simultaneously observing the behavior of several models of test elements (test specimens, test tubes) in time and by studying how these properties change at the relevant indicators point by point. The behavior and state of each variable and its movement in time and space are analyzed, on the basis of which the function of the expected result is compiled. The unification of the individual elements is most often carried out by integrating the results by the relevant components and indicators.

3. Review of mathematical statistics and probability theory

An event is a quantitative or qualitative result of a test under clearly defined conditions. An event is when a steel test specimen (tube) breaks under a certain load, and similarly, when it does not break under another load. If an event always occurs under a given set of conditions, it is called probable; if it never occurs, it is called impossible [66-71]. If an event may or may not occur under certain conditions, it is called conditional. The object of study of probability theory is random events [67-102]. Initially, events occur or do not occur under the same conditions. By the same set of conditions in each experiment, we understand a controllable set of conditions, as long as the experiment is controllable. In the case of random events, this means that our knowledge of the phenomenon is insufficient, that there are unexplained factors that influence the outcome of the experiment, or that the studied factors have no influence, as far as our knowledge takes them into account. What must always be added is that knowledge of the main characteristics of the phenomenon, of the main factors determining the phenomenon, is a prerequisite, and the unknown is only a kind of correction of the known. In other words, probabilistic methods cannot be applied directly to completely unknown and unexplained phenomena. In other words, it is not essential to assume that a given event is random. In the example of the steel specimen (tested in tension) mentioned above, it is not enough to say that failure in the tensile stress range of 450 - 460 MPa and failure in the range of 470 - 480 MPa are random events. It is clear that for a given grade of steel these two random events actually have different probabilities of occurrence. The concept of probability is fundamental to probability theory and is used in many areas of science where this theory is applied [67-78]. It is not easy to define this concept perfectly simultaneously from the point of view of mathematics, philosophy and applied sciences. Therefore, if n is the number of trials in a given sequence of trials, and m is the number of favorable trials in this sequence, i.e. the number of trials in which the random event in question occurs [65]. From this one $\frac{n}{m}$ dependence gives the frequency of occurrence of the random event. It is clear that this frequency is an empirical characteristic of the event [66]. With the continuous increase in the number of trials n, the frequency approaches a certain value and is seen to vary continuously around this value [66-81]. This value is the theoretical probability, and the fact that the frequency approaches the probability with increasing number of trials reflects the fundamental law of large numbers [66]. It is important to emphasize again that probability is a theoretical property and concept of probability theory, while frequency is an empirical property and concept of mathematical statistics. The probability of a random event is a precisely defined quantity, while frequency is variable [66]. If we consider an experiment in which either event A, event Б, or both events do not occur simultaneously. Such a conditional event is called a discrepancy: what is the probability that one of the two events, i.e. A or Б, will occur? [75-76]. The number of random occurrences of event A is denoted by $\mathbf{n_A}$, and the number of random occurrences of event Б is denoted by $\mathbf{n_Б}$, and $\mathbf{n}$ denotes the number of total trials, then the frequency of event A+Б, which means that either event A or event B is realized, is determined (for.1.3.1):

$$\frac{n_A + n_Б}{n} = \frac{n_A}{n} + \frac{n_Б}{n} \tag{1.3.1}$$

Using the theory of large numbers, we obtain the probability addition theorem (for. 1.3.2):

$$P(A + Б) = P(A) + P(Б) \tag{1.3.2}$$

If we have a situation in which two random events occur, one of which (A) is realized only if the second (Б) is realized, while the occurrence of event Б is not related to event A [66]. We denote by $\mathbf{n_A}$,

$\mathbf{n_Б}$ and $\mathbf{n_{АБ}}$ respectively the number of cases favorable for A, for Б and for both events simultaneously, and with the n-total number of cases. Because in the situation under consideration $\mathbf{n_A = n_{АБ}}$, to (for.1.3.3):

$$\frac{n_{АБ}}{n} = \frac{n_А}{n_Б} . \frac{n_Б}{n} \tag{1.3.3}$$

Then the probability is determined by formula 1.3.4:

$$P(АБ) = P(А|Б) . P(Б) \tag{1.3.4}$$

where: **P (АБ)** is the probability of both events occurring simultaneously **A** and **Б**, a **P(A|Б)** is the probability of the event occurring **A** provided that the event has occurred **Б** (also known as "conditional probability").

For two events **A** and **Б** are independent if the occurrence of one does not depend on the occurrence of the other. For such events **P(A|Б) = P(A)** and from for.1.3.4 follows:

$$P(АБ) = P(А) . P(Б) \tag{1.3.5}$$

Formula (for.1.3.5) shows the relationship also known as the probability multiplication theorem.

From a practical point of view, the greatest benefit is a random event expressed in the measurement of a given quantity (I will denote it by **X**), for which all values in the interval $(-\infty ; +\infty)$ exist. This quantity will be a random quantity [66-77]. It can take on all possible values, but the realization of a certain value of it is not equally likely. For complete information on the random quantity **X**, a distribution function **F(X)** is used, which is equal to the probability of realizing a value of **X < x** (fig.1.3.1), according to the formula [66-77]:

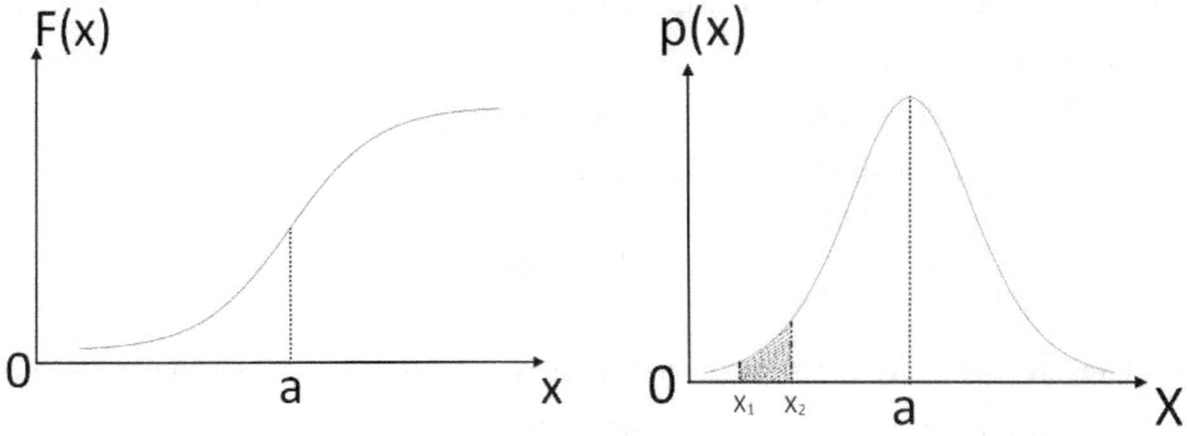

(a) (b)

Figure 1.3.1 (a) Graph of a monotonically increasing function; (b) Probability density graph of the distribution function

$$F(X) = P(X < x) \tag{1.3.6}$$

From the definition it follows that **F (-∞) = 0**, and **F (+∞) = 1**, and in the interval (-∞ ; +∞) the function is non-decreasing [66-77]. In the case that the random variable X takes all values from the number axis, then **F(X)** is a monotonically increasing function and is a step function (fig.1.3.2), and if the random variable **X** takes discrete values, i.e. $\mathbf{x_1}$, $\mathbf{x_2}$, $\mathbf{x_3}$ and so on, then from the first derivative of the function a probability density is determined, which in probability theory is fundamental (for.1.3.7) [66-77]:

$$p(x) = \frac{dF(x)}{dx} \tag{1.3.7}$$

The probability density graph of the distribution function is the classical Gaussian distribution and is shown in fig.1.3.1(b) [66-77].

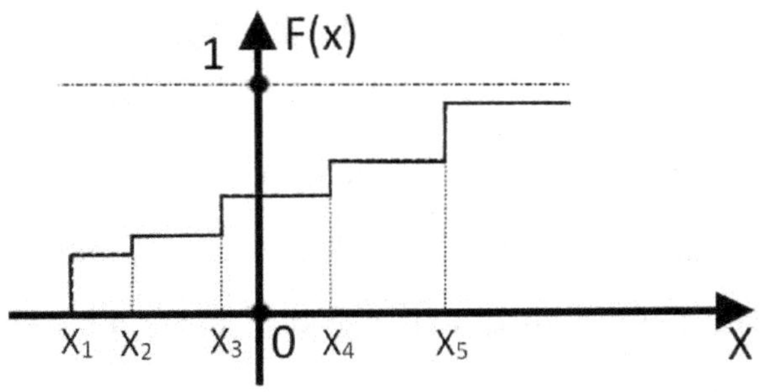

Figure 1.3.2 Stepwise growing function

The mode is the maximum value of the abscissa for **p(x)**, and for the case shown the mode is **x = a**. The probability that the value of X will be in the infinitely small interval **(x, x+dx)** using the for.1.3.6 and for.1.3.7 is [66-83]:

$$P(x < X < x + dx) = dF(x) = p(x)dx \tag{1.3.8}$$

Respectively, the probability is in a finite interval $\mathbf{x_1 < X < x_2}$ is:

$$P(x_1 < X < x_2) = \int_{x_1}^{x_2} p(x)dx \tag{1.3.9}$$

Therefore, the probability that the random variable **X** falls into a given interval is numerically equal to the area between the probability density graph in this interval and the horizontal axis. For the value of **X** to be in the interval **(-∞ ; +∞)** to be a certain event, then the entire area enclosed by the graph must be equal to **1** [66-77], i.e.:

$$\int_{-\infty}^{+\infty} p(x)dx = 1 \quad (1.3.10)$$

The two functional characteristics, **F(X)** and **p(X)** are carriers of the necessary and sufficient data for the random variable **X**. Considering the meaning of formula (for.1.3.7), these are the integral and differential distribution function. Having one of these functions, we are able to find the other function, using formula [66-77]:

$$F(x) = \int_{-\infty}^{+\infty} p(x)dx \quad (1.3.11)$$

Finding these functions is not necessary, it is enough for the given random variable **X** to know some numerical characteristics. In order to know these characteristics, we should make n number of measurements of the random variable **X**, as in n_1 case, we will have x_1, n_2 case we will have x_2 and so, by generalizing we get that for nm case we will have x_m values, then the arithmetic mean value is determined by the formula [66-77]:

$$\overline{X} = \frac{1}{n}\sum_{k=1}^{m} x_k n_k = \sum_{k=1}^{m} x_k \frac{n_k}{n} \quad (1.3.12)$$

From the attitude $\frac{n_m}{n}$ moving on to probabilities $\mathbf{P_m}$, it follows that there will be a quantity analogous to the empirically obtained arithmetic mean value, which is the theoretical mean value [66]. In the mathematical literature [66-102] it is defined as the mathematical expectation of the random variable **X**, and is denoted in a variety of ways. The formula for the direct distribution is [66-77]:

$$\langle X \rangle = \sum_{k=1}^{m} x_k P_k \quad (1.3.13)$$

In the case of a continuous distribution of the random variable, it will pass into an integral according to the formula (for.1.3.8.) [66-77]:

$$\langle X \rangle = \sum_{-\infty}^{+\infty} x\, p(x)dx \quad (1.3.14)$$

In fact, the mathematical expectation shows the most probable value of the random variable. An interesting fact is that this variable does not coincide at all with the so-called mode and if we look at the formula (for.1.3.14), then $\langle \mathbf{X} \rangle$ coincides with the static moment of the area of the diagram **p(x)** relative to its ordinate axis [66-77].

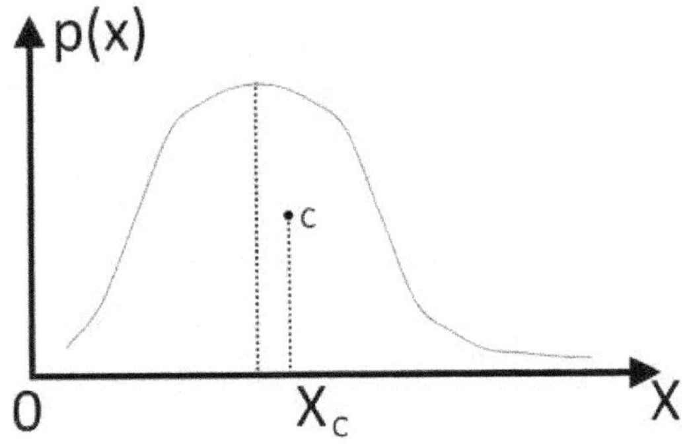

Figure 1.3.3 Visualizing the center of gravity of the function **p(x)**

Considering formula (for.1.3.10), it turns out that this area is equal to 1, it follows that $\langle \mathbf{X} \rangle$ gives the abscissa of the center of gravity of the area (fig.1.3.3) composed of the function **p(x)**, and it will coincide with the so-called mode only in the presence of symmetry in the graph of the function [66-77].

Using the classical theory of the resistance of materials, since there is a given area (even from a diagram), there are also moments of inertia and moments of higher order, which can be determined by the formulas [66-77]:

$$\langle X^2 \rangle = \int_{-\infty}^{+\infty} x^2 \, p(x) dx \tag{1.3.15}$$

$$\langle X^n \rangle = \int_{-\infty}^{+\infty} x^n \, p(x) dx \text{ , при } n > 2 \tag{1.3.16}$$

The second moment of area is obtained directly by reasoning according to the theory of mechanics of materials (for.1.3.15), but the interesting thing is that the mean square of the random variable **X** is obtained (for.1.3.13 and for.1.3.14) [66-77]. It is similar for the formula (for.1.3.16) for the next moments of higher order [66-77].

The dispersion or also called the centered mean square (denoted $\langle \mathbf{X^2} \rangle$) is calculated by calculating the mean square at the ordinate axis passing through the center of gravity of the **p(x)** function, i.e. [66-77]:

$$D(X) = \langle (X - \langle X \rangle)^2 \rangle = \int_{-\infty}^{+\infty} (x - \langle X \rangle)^2 . \, p(x) dx \tag{1.3.17}$$

The dispersion of a random variable around its mean value is practically its variance [66-77]. The clustering of given values around an expected value represents insignificant dispersion or minimal deviation from the value. In this case, the function **p(x)** has a clearly pronounced maximum and quickly falls off after it. The term standard deviation (standard **σ**) of the random variable is usually used, and it is according to the formula [66-77]:

$$\sigma = \sqrt{D(X)} \tag{1.3.18}$$

A distinction should be made between the input quantities and characteristics – the functional **F(x)** and **p(x)** and the numerical ones ⟨⟨**X**⟩ and ⟨$\mathbf{X^2}$⟩. The functions **F(x)** and **p(x)** contain all the necessary data about the probabilistic nature of the random variable **X**, while the numerical ones actually promise certain data about this random variable. From the classical theory of resistance of materials, it has been established that two figures can have the same area, the same center of gravity and even the same moment of inertia, but be completely different (different geometry) [66].

In Gaussian distribution for the normal probability distribution, the formula used is [65-93]:

$$p(x) = \frac{1}{\sigma\sqrt{2\pi}} \, . \, e^{-\frac{(x-a)^2}{2\sigma^2}} \tag{1.3.19}$$

The condition given in formula (for.1.3.10) is fulfilled for the function of for.1.3.19 and for this function the dependence [65-93] is still valid:

$$\int_{-\infty}^{+\infty} x\, p(x)dx = a \; , \; \int_{-\infty}^{+\infty} (x-a)^2\, p(x)dx = \sigma^2 \tag{1.3.20}$$

From which it follows that the parameter a is the mathematical expectation, and the parameter **σ** is the mean square deviation (standard) [66-77]. The probability density of the normal distribution is due to the differences in the standard values of **σ**. As mentioned above, the smaller the standard value **σ** of a random variable, the smaller the dispersion around its mean value (mathematical expectation), which coincides with the mode in Gauss's rule due to symmetry [66-77]. The normal (Gaussian) distribution is widely used in probability theory and mathematical statistics, as well as in mechanics. In particular, the following fact: if a random variable has a normal distribution, its distribution is preserved after linear transformations (including differentiation and integration) [66-77].

However, there are other factors. Normalization is a common phenomenon in the world around us. The need for it is justified by the idea of ignoring the dispersion or deviation from the base, random errors in various types of measurements (geodetic, laboratory, etc.). According to the theorem, the normal distribution (or a distribution close to it) is given by a random variable, which is the result of a series of independent causes [68].

In the literature [66-77], the concept of an n-dimensional joint distribution function is introduced, which is used when several random variables should be summarized as a set. According to the formula [66-77]:

$$P\begin{pmatrix} X_k < x_k \\ k = 1,2,\dots\dots\dots\dots, n \end{pmatrix} = F(x_1, x_2, \dots\dots\dots\dots\dots\dots\dots\dots, x_n) \tag{1.3.21}$$

In turn, the n-dimensional joint probability density is given by the formula [65-77]:

$$p(x_1, x_2, \dots\dots\dots\dots, x_n) = \lim_{\Delta x_k \to 0} \frac{P\begin{pmatrix} x_k < X_k < x_k + \Delta x_k \\ k=1,2,3\dots\dots n \end{pmatrix}}{\Delta x_1 . \Delta x_2 . \Delta x_3 \dots\dots\dots . \Delta x_n} \tag{1.3.22}$$

When applying the method of analogy, it is established that there is a relationship between the two formulas (for.1.3.21 and for.1.3.22) [65-77]:

$$p(x_1, x_2, \dots\dots\dots\dots, x_n) = \frac{\partial^n F(x_1, x_2, x_3, \dots\dots\dots\dots\dots x_n)}{\partial x_1\ \partial x_2\ \partial x_3 \dots\dots\dots\dots . \partial x_n} \tag{1.3.23}$$

$$F(x_1, x_2, \dots\dots\dots\dots, x_n) = \int_{-\infty}^{x_1} \int_{-\infty}^{x_2} \int_{-\infty}^{x_3} \dots\dots\dots\dots \int_{-\infty}^{x_n} p(x_1, \dots\dots . . x_n)\, dx_1 \dots\dots dx_n \tag{1.3.24}$$

According to formula (form.1.3.24) there exists the property (as in form.1.3.10) [65-77]:

$$\int_{-\infty}^{+\infty} \int_{-\infty}^{+\infty} \int_{-\infty}^{+\infty} \dots\dots\dots\dots\dots \int_{-\infty}^{+\infty} p(x_1, \dots\dots . . x_n)\, dx_1 \dots\dots dx_n = 1 \tag{1.3.25}$$

Based on the probability multiplication theorem (for.1.3.5), it is established that if the random variables are independent, then the joint probability density is equal to the product of the densities of each of them, i.e. [65-77]:

$$p(x_1, x_2, \dots\dots\dots\dots, x_n) = p(x_1)\, p(x_2)\, p(x_3) \dots\dots\dots\dots\dots p(x_n) \tag{1.3.26}$$

Respectively, for the set of several random variables, we can introduce (numerical) characteristics of the distribution or the so-called mean average value [66-77]:

$$\langle X_k \rangle = \int_{-\infty}^{+\infty} x_k\, p(x_k) dx_k \tag{1.3.27}$$

For second-order moments [65-77]:

$$\langle X_j X_k \rangle = \iint\limits_{-\infty}^{+\infty} x_j x_k \, p(x_j ; x_k) dx_j \, dx_k \qquad (1.3.28)$$

For third-order moments [65-77]:

$$\langle X_i X_j X_k \rangle = \iiint\limits_{-\infty}^{+\infty} x_i x_j x_k \, p(x_i ; x_j ; x_k) dx_i \, dx_j \, dx_k \qquad (1.3.29)$$

Using the method of analogy (for.1.3.17), the centered second-order moments are also introduced according to the formula [65-77]:

$$K_{x_j x_k} = \langle [(X_j - \langle X_j \rangle)(X_k - \langle X_k \rangle)] \rangle = \\ = \iint\limits_{-\infty}^{+\infty} (X_j - \langle X_j \rangle)(X_k - \langle X_k \rangle) \, p(x_j ; x_k) \, dx_j \, dx_k \qquad (1.3.30)$$

$$K_{x_j x_k} = \int\limits_{-\infty}^{+\infty} (x_j - \langle X_j \rangle \, p(x_j) dx_j \, . \int\limits_{-\infty}^{+\infty} (x_k - \langle X_k \rangle \, p(x_k) dx_k = \\ = \left[\int\limits_{-\infty}^{+\infty} x_j \, p(x_j) dx_j - \langle X_j \rangle . \int\limits_{-\infty}^{+\infty} p(x_j) dx_j \right] . \left[\int\limits_{-\infty}^{+\infty} x_k \, p(x_k) dx_k - \langle X_k \rangle . \int\limits_{-\infty}^{+\infty} p(x_k) dx_k \right] \qquad (1.3.31)$$

These moments reflect the correlation (statistical dependence) between the random variables X_j and X_k. Uncorrelated are those random variables whose correlation moment is equal to zero (using for.1.3.26 and for.1.3.30). A random variable **X** can be a variable in some function **Y=F(X)**, then this function will be a random function and its probability properties will depend on the properties of the random variable **X**. The formula by which the separation function of **Y=F(X)** can be found is [65-77]:

$$F(y) = P(Y < y) = P \, [f(X) < y] \qquad (1.3.32)$$

Using formula (for.1.3.9) we obtain [65-77]:

$$F \, (y) = \int\limits_{(A)} p(x) dx \qquad (1.3.33)$$

where **(A)** is conventionally denoted by all sets of intervals on the abscissa axis for which the condition **F(X) < y** is satisfied. In case the function has one interval, then it is monotone, i.e. [65-77]:

$$F(y) = \int_{-\infty}^{y} p(x)dx \tag{1.3.34}$$

This is a special case, but it means that an inverse function exists **x= φ(y)** and the probability density is determined by the formula (using formula1.3.24 dx c $|\varphi'(y)|$) [65-77]:

$$F(y) = \int_{-\infty}^{y} p[\varphi(y)].\,|\varphi'(y)|dy \tag{1.3.35}$$

or after processing (using formula 1.3.7), we finally obtain [65-77]:

$$p(y) = p[\varphi(y)].\,|\varphi'(\vartheta)| \tag{1.3.36}$$

Since the probabilities for the entire interval should be summed, the requirement for an absolute value is set **(- ∞; y)**, including also sub-intervals where the conditions are met $\boldsymbol{\varphi' < 0}$ [65-77].

The main priority of mathematical statistics is to find the theoretical probability distribution of the random event of interest, given the empirical frequency distribution. In this part, the empirical distribution is assumed to be given, but in practice it must be constructed a priori by statistical processing of the initial data. In the practical aspect, these are n measurements of the random variable **X**, as a result of which a series of values x_1 , x_2 ,........., x_n (variation order) is obtained. The values are grouped into intervals **dx (ΔX)** of equal size (step). These intervals should be sufficient in number to smoothly track the change of the random variable [66-77]. The next step is to calculate the average relative frequencies for each of the intervals according to the formula [65-77]:

$$\bar{n}_i = \frac{1}{\Delta x} \cdot \frac{n_i}{n} \tag{1.3.37}$$

where: $\mathbf{n_i}$ – number of cases in **i**-th interval; **n** – the total number of cases.

The histogram (fig.1.3.4(a)) is obtained by plotting the quantities as ordinates. Due to the fact that the empirical probability density curve (fig.1.3.4(b)) corresponds to a probability theoretical curve, the step area must be equal to **1** [65-77]. This is because in the formula the denominator is **X**. We can also outline an empirical polygon whose vertices are in the middle of the intervals **ΔX** and draw an empirical distribution curve (cumulative curve), for each ordinate the condition must be met that it is equal to the areas of the part of the histogram existing to the left of this ordinate [65-77]. Using the variation order, basic numerical characteristics of the empirical distribution are calculated, such as arithmetic mean and square root [65-77]:

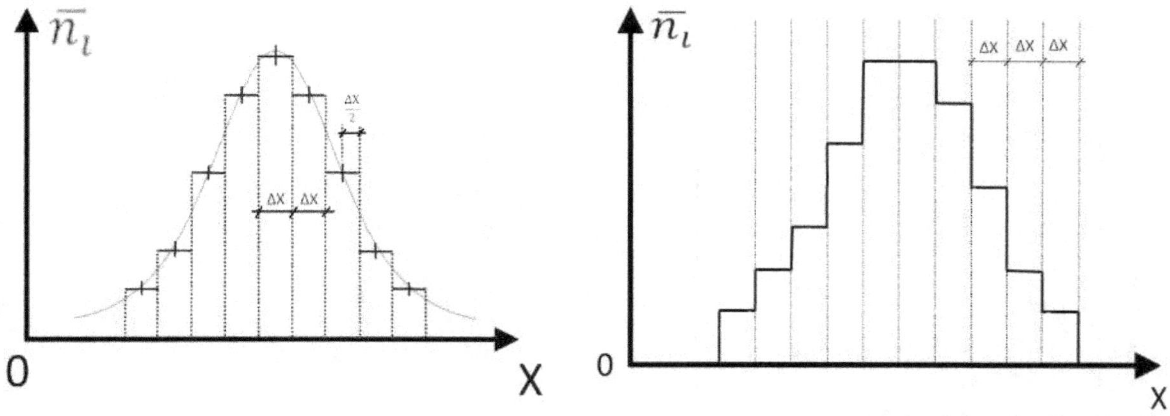

(a) (b)

Figure 1.3.4 (a) Histogram represented by an empirical polygon; (b) Empirical probability density curve

$$\overline{X} = \frac{1}{n}\sum_{k=1}^{n} x_k \tag{1.3.38}$$

$$\overline{(x - \overline{X})^2} = \frac{1}{n}\sum_{k=1}^{n} (x_k - \overline{X})^2 \tag{1.3.39}$$

The empirical analogue of these two quantities is the mathematical expectation **⟨X⟩** and the variance **D(X)**. The empirical analogue of the standard σ is closer to the mathematical expectation and is found by the formula [65-77]:

$$s^2 = \frac{1}{n-1}\sum_{k=1}^{n} (x_k - \overline{X})^2 \approx D(X) \tag{1.3.40}$$

It serves as a measure of the dispersion of the measured values. The coefficient of variation (change) is their ratio.

From a practical point of view, for a chosen theoretical probability density, we construct a distribution function **F(X)** (fig.1.3.5(a)) and compare it with a corresponding empirical polygon $\mathbf{\overline{F}(x)}$ (fig.1.3.5(b)), after which we find the maximum absolute value difference by the formula [65-77]:

$$D_n = \max|F(x) - \overline{F}(x)| \tag{1.3.41}$$

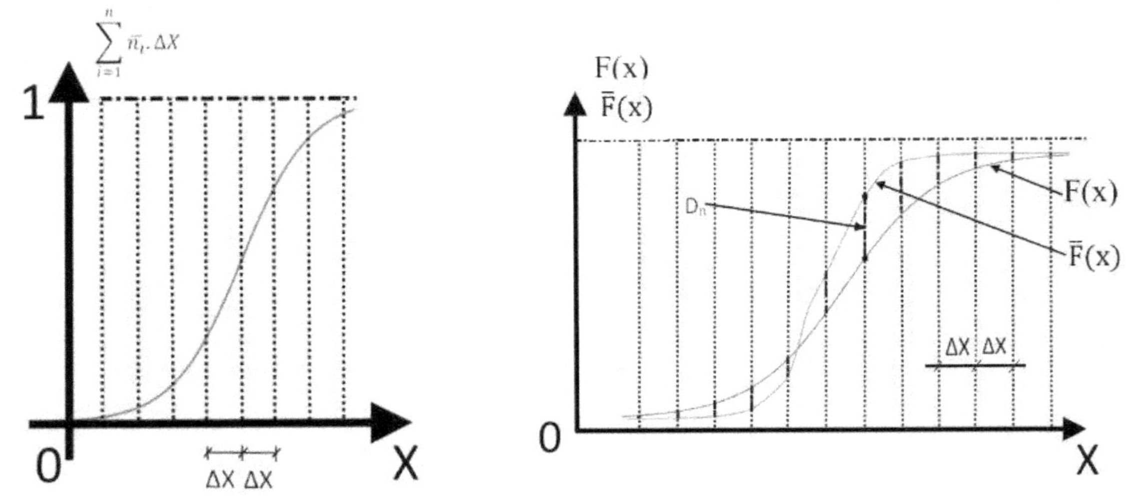

(a) (b)

Figure 1.3.5 (a) Empirical distribution curve; (b) Distribution function and comparison with a corresponding empirical polygon

Found on the **n** measurements made. Then we look for the probability $\mathbf{P(D_n)}$, since in the result of the measurements of the random variables there is a difference of size $\mathbf{D_n}$ and we determine it by the formula [65-66]:

$$P(D_n) = 1 - K(D_n\sqrt{n}) \tag{1.3.42}$$

where:

$$K(x) = \sum_{a=-\infty}^{+\infty} (-1)^a . e^{-2.a^2.x^2}$$

The above is known as the Kolmogorov agreement criterion [65-66].

4. Review of nonlinear time trending

Predicting the future state of a process or one or more of its components depending on various factors that develop over time has been a task for decades, especially in the natural sciences [6]. In the case that a process is described by a given mathematical model (by some function), its development represents a trend over time and from the possible mathematical solutions taking into account the initial conditions at $\mathbf{t = 0}$, leads to possible predictions at some future moments in time [6-12]. In this assumption, the establishment of such functions should be error-free and this is transformed into a formula or equation of the process, since it is true for any arbitrary $\mathbf{t > 0}$ (time progresses) and the predicted value is actual and real [6-12]. In the case of corrosion, the factors that determine the process are a sufficient number of unknown components. This assumption cannot be accepted as authoritative and the deviation (permitted error) is significant. We can assume that the process is linear, then these errors are transformed in time without change, as a result of which an absolutely unreliable result is obtained and the practical and theoretical meaning of the need to establish it is lost [1-12]. Under the assumption that the process is nonlinear, it automatically means that it is very sensitive to its initial conditions, the factors of change with the passage of time and, above all, to the accepted hypothetical assumptions for its development in time [1-5]. A deterministic trend for a given period **T** based on some function **x(t)** consisting of more than one characteristic of the state of the process, including its initial state, should be described by the formula [1-9]:

$$x(t+T) = F_t(x(t)) \tag{1.4.1}$$

and when the process is continuous [1-9]:

$$X_{t+T} = F_t(X_n, X_{n-1}, X_{n-2} \dots \dots \dots \dots X_{n-(n-1)}) \tag{1.4.2}$$

The main idea is to construct (graphically construct) a predictive function $\mathbf{F_t(x)}$. This type of function can be local, which means that for each step it should be different and separate in order to obtain the nonlinear function (as a whole), even under the condition that the individual elements are linear [8-9]. In the presence of a global function, it automatically spreads over all existing data and can be used forward in time, i.e. the possibility of a so-called special case (private solution) is excluded [8-9].

For the local approach, a linear approximation is used [14]. It is based on Lyapunov's theory [18-19], according to which the local mechanics of mechanical processes for any deterministic system is approximately linear in a given interval (tending to zero or **dx**) and thus a global nonlinear dynamic of the system is achieved. In general, the state at the initial time t and the future time **t+T** are points in the time theory, located at different levels in time and are connected by a predicted theory of nearby neighboring points in the reconstructed phase space [5]. For them, there are certain coefficients of the local approximations found by the method of least squares [5]. In case it is presented as a polynomial of higher degree, the accuracy of the trend is significantly improved [14-15].

For the global approach, the approximation can be from: polynomials, rational functions, Fourier series, neural networks, spectral number series and other nonlinear methods of higher mathematics [14-15].

In case of weak nonlinearity, a suitable method is the autoregressive model method, according to the formula [104-112]:

$$X_n = X_0 + \sum_{i=1}^{m} a_i . X_{n-i} + e_n \quad (1.4.3)$$

Using the formula above, a generalized model of the process with improved forecast values is created - the accuracy increases, the calculation error decreases significantly (the author understands that a significant reduction or increase in the error is when the accumulated error increases/decreases by more than 25%).

Practice has imposed the opinion that the forecast value **X** will not coincide with the true value **X**, i.e. there is always an error or deviation, which leads to the fact that the forecast value is not perfect nor accurate (precisely calculated) and contains an error/deviation that should be within the permissible range. For this reason, quantitative estimates of the forecast error (allowed error) are made according to the formula [104-112]:

$$PG = \sqrt{\frac{\sum_n (X_1 - X_n)^2}{\sum_n (X_n - \bar{X})^2}} \quad (1.4.4)$$

From mentioned at value of **PG=0** cannot exist, and the maximum value is **PG = 1** (in ideally) and this means **0 < PG < 1**.

The simple linear correlation coefficient **R(T)** between **X** and $\bar{\mathbf{X}}$ is used to establish the predictive qualities of the constructed computational model.

On the other hand, any measurement of a particular parameter or group of parameters as a result of an experiment implies that there is uncertainty in the measurement. This means that there is a parameter attached to the measurement result that characterizes the dispersion of the values, and that they can be reasonably attributed to the measurand. Uncertainty arises from different sources and is categorized in different ways. Usually, for the uncertainty of the effect of corrosion on the working diagram of steel, six main points are identified that are important for the assessment of uncertainty. The modeling approach can use these effects as the basis for the mathematical model. Alternatively, classical theory identifies six different sources of error in the investigation of the result, each of which can be reduced to different cause-effect factors, which in turn can be used in different modeling approaches. Another alternative approach is to consider all steps in the data collection process as sources of uncertainty that contribute to the uncertainty of the final result. The simplest research model considers uncertainty as arising from four domains, and the domains are considered as sources of uncertainty in the stochastic model, which is consistent with the grouping of sources of uncertainty.

Establishing the trend of a given process through time extrapolation is practically an impossible solution, but by reconstructing the phase space it is possible to replace the extrapolation with a corresponding type of interpolation that will provide results with sufficient accuracy for practice.

5. Review of the stress-strain diagram of steel

For each steel element there is a σ-ε diagram. This diagrammatic dependence, according to different scientists, has a different name - a stress-strain diagram of the material, a deformation diagram, a stress-strain curve, etc. In all its variations of name, it essentially represents a visualization of the **σ-ε** dependence, with the deformation **(ε)** on the abscissa and the stress **(σ)** on the ordinate. Most often, the dimension of the deformation is [mm], and of the stress [MPa or kN/cm2]. Some of the first data on the existence of this diagram date back to 1896. Since then, the diagram has been the subject of various studies. It is these studies that have contributed to the creation of dimensional theories, strength theory, standards and other documents. Depending on the way in which the data from the diagram are treated, the corresponding concepts are established and defined.

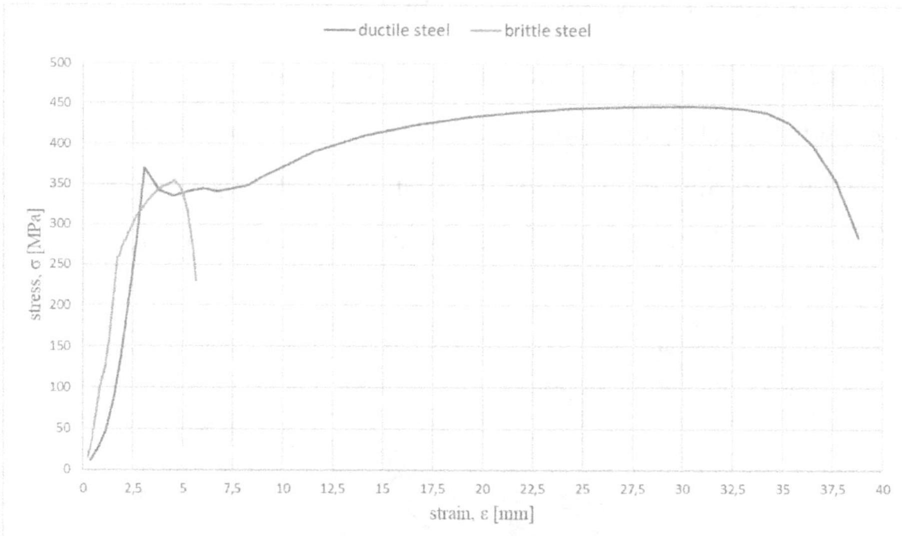

Figure 1.5.1 Steel stress-strain diagram – ductile and brittle

One of them is a ductile (tough) or brittle material – **σ-ε diagram** of ductile and brittle steel is shown in fig.1.5.1 [113]. In ductile material (fig.1.5.1) there is a clearly expressed (visible) yield point and the material breaks after the presence of plasticization and very large deformations (for steel S355JR it is of the order of 35 – 45 mm), brittle material (fig.1.5.1) is one in which the yield point almost does not exist (it is too small, difficult to notice) the term "no yield point" is usually used and the failure occurs at very small deformations (for high-strength steel Docol 1400M it is of the order of 3-8 mm), which leads to instantaneous failure.

Fig.1.5.2(a) and fig.1.5.2(b) show the visible difference in the material after failure (in the section where it occurred). Before failure occurs in ductile materials, a so-called neck is formed in which the crack initiation occurs, in brittle materials this effect is absent immediately before failure and the crack formation leads to its rapid growth and sudden failure occurs [113].

(a) (б)

Figure 1.5.2 Demolition section view – (a) brittle; (б) ductile [112]

To obtain the σ-ε diagram, this is done by observing a relevant standard – a test body (with certain dimensions) and a tensile testing machine. The test body, the so-called test tube, is placed in the machine, which begins to test it in tension by gradually increasing the force by a corresponding step and determines the corresponding values of the deformation at the corresponding stress.

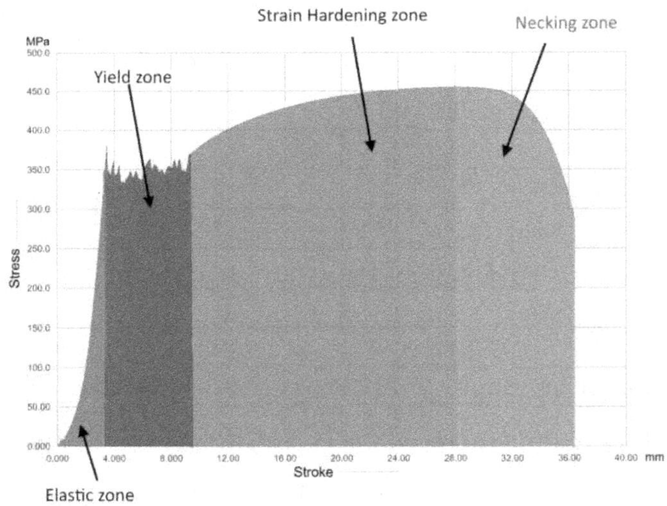

Figure 1.5.3 Stress-strain diagram divided into zones (areas)

The stress-strain curve of steel is divided into 4 zones (areas) (fig.1.5.3). Each of these zones (areas) is characterized by corresponding mechanical properties that are relevant to the relevant dimensional or structural calculations.

In turn, this diagram also determines the main points in it, determining the change in the mechanical properties of the material shown in fig.1.5.4. For steel, point A of the diagram is called the "proportionality limit", up to this point Hooke's law for the linear relationship between stresses and strains is in effect. The E-modulus, also called the modulus of elasticity, is a material quantity (which is constant) and is defined as the stress (normal) σ to the change in the relative longitudinal strain ε, according to Hooke's law. The next point on the diagram (point B) is known as the "elastic limit", and its characteristic is that if we stop loading up to this point, then the tested steel (tube) will not have any residual deformations, i.e. the behavior of the material is completely elastic. The next characteristic value is pointing C – this is the so-called "upper yield strength". This point is characterized by the fact that after it the deformation begins to increase without a significant increase in stress, i.e. the plasticization process has begun. If after this point we stop the loading, then plastic deformations will remain in the steel (tube). Somewhere after point C and the end of the yielding zone is point D, which is the "lower yield limit" and determines the minimum value of the stress at which irreversible plastic deformations have occurred. In turn, point E represents the maximum strength (tensile) – this is the maximum value of the stress that the test specimen can withstand after the start of its plasticization. The last is point F which is the "failure point", at which the deformations are maximum, but the stress at whose value the failure (separation) occurs is much lower than the maximum established strength [113].

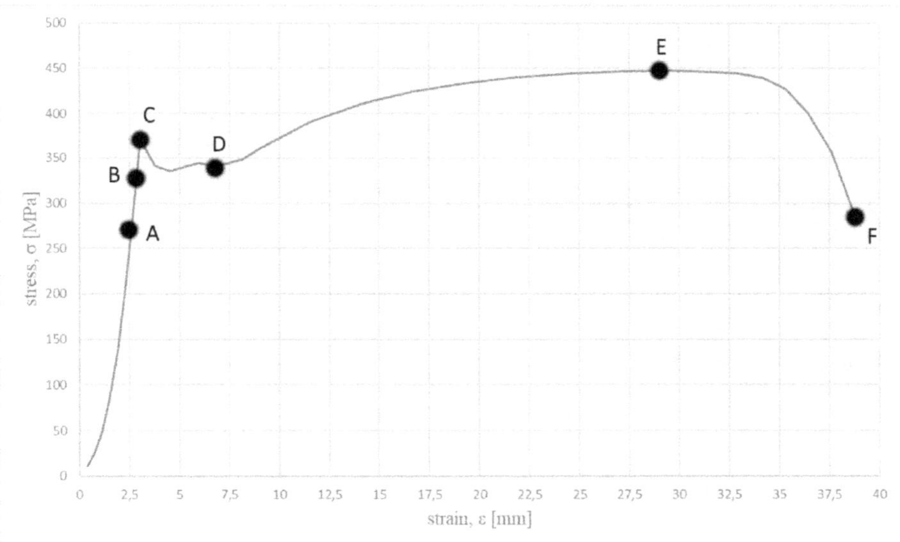

Figure 1.5.4 Stress-strain diagram - basic points

6. Summary

The short reviews on statistical mechanics, mathematics of statistics, probability theory, nonlinear determination of dependencies and the steel stress-strain diagram with the corresponding formulas mentioned in this chapter clearly state the existence of a theoretical approach for establishing a corresponding trend based on a sufficient number of results. The application of this theoretical approach in a variety of areas of existing engineering problems should lead to an approximate solution that should be used in practice. The ability to establish the change in the main areas or points of the steel stress-strain diagram over time definitely optimizes the process, leading to maximum operational use of steel as a material in building structures and facilities. The practical focus is expressed in establishing a polynomial dependence in the form of a function for the change in the mechanical properties of steel with corrosion depending on the time of the corrosion influence.

CHAPTER II. BASIC PRINCIPLES

1. Principle of brittle failure according stochastic theory

There are two types of steel failure depending on its chemical composition, structure and method of production. Ductile (tough) steel is characterized by the fact that before failure occurs, the steel element undergoes significant deformations (tolerates plastic deformations), while other types of steel are characterized by brittle failure (almost no plastic zone). Applied mechanics considers the failure of steel, but this is an extremely difficult process from the non-probabilistic (deterministic) aspect. Some authors [35] believe that an acceptable solution to this problem cannot be achieved without using a probabilistic approach. We assume that this assumption is true, since any corroded steel is characterized not by a continuous structure, but by a fairly large number of initial volumes, i.e. a crystal lattice (in metals), which is not uniform [116]. These properties and the characteristics of the contact zone between the structures will form the properties of the corroded steel in a statistical manner. Destruction is also closely related to the ability to find the primary volumes or low-density contact zones.

The principle governing the interactions between the basic elements in corroded steel (volume, contact area, structure) are not well understood and, furthermore, the processes occurring in corroded steel up to the point of failure are irreversible. If a comprehensive probabilistic theory of deformation and failure is developed that encompasses brittle failure and plastic deformation in corroded steel, this will provide the basis for further development and understanding of the processes.

From the reviews done in the theory of brittle fracture [57] it is established that practically applicable principles and methods are set out in the Waloddi Weibull theory (Weibull distribution from 1939), upgraded by Yakov Frenkel and Tatiana Kontorova (Frenkel–Kontorova model from 1938) [65-66]. These authors have proposed initially that there should be a first approximation, by neglecting the influence of the relevant dependencies between the primary elements, provided that the strength of the steel will be determined by the strength of the weakest link (in it) among them. The principle is that a uniform stressed state with equivalent stress **S** is realized in the steel, and by $\mathbf{F_r}$ we denote the value of the failure stress **S**. We introduce the value **R** for a brittle material and this is the failure limit, to it we also introduce a distribution function **f(R)** which reflects the probability of defects in the steel that cause local failure less than the value **S**. If by **n** we denote the average number of defects per unit volume, it follows that in volume **V** we have **n.V** corrosion defects. It logically follows that the probability $\mathbf{f_v(R)}$ that there exists a place in the steel where the condition $\mathbf{R < S}$ is fulfilled as a result of **n.V** defects and then for the minimum values of the distribution function $\mathbf{f_{n,min}(x)}$, taking into account that the process is random, is according to the formula [65-66]:

$$f_v(R) = 1 - (1 - f(R))^{n.V} \tag{2.1.1}$$

Certainly, the number **n.V** is quite large (in numerical terms); therefore, the asymptotic formula [65-66] is applied, i.e.

$$f_v(R) \approx 1 - e^{[-c.n.V.(R-s_0)^{\alpha}]} \tag{2.1.2}$$

where: s_0 – reduced stress on the weakest part. α and **c** – not certain constants (of this step).

We use a known reference volume V_0 and an undefined constant for this reference volume s_c, for the function which is according to form 2.1.2, namely [65-66]:

$$c.n = \frac{1}{V_0 . s_c^{\alpha}} \tag{2.1.3}$$

After making the substitution in the formula (form.2.1.2), we obtain [65-66]:

$$f_v(R) = 1 - e^{\left[-\frac{V}{V_0}.\left(\frac{R-s_0}{s_c}\right)^{\alpha}\right]} \tag{2.1.4}$$

For the undefined constant s_c, it is found that it has a stress dimension i.e. MPa or kN/cm^2.

In the general case of the problem, the stress state is non-uniform, i.e. the main normal at different points of the element have different orientations. The volume of each element can always be divided into smaller volumes (parts) – the volume **V** of the steel (element) is divided into a finite number of small volumes ΔV_k, which are so small that with sufficient accuracy for practice we can assume that for each of them the stress state is uniform, but at the same time they are large enough to apply the formula (form.2.1.4) directly. In this case P_v (the probability of destruction of the entire volume) according to the theory of probability is equal to the sum of the probabilities of the small volumes ΔV_k, which we write by the formula [65-66]:

$$P_v = e^{\left[-\frac{1}{V_0}.\Sigma_k \Delta V_k.\left(\frac{S_k-s_0}{s_c}\right)^{\alpha}\right]} \tag{2.1.5}$$

Then the probability of failure of the element with volume **V** can be calculated using the formula:

$$P_v^* = 1 - P_v \tag{2.1.6}$$

Applying mathematical statistics, we find the mathematical expectation using the formula:

$$P_v(R) = \frac{dF_v(R)}{dR} = \frac{V}{V_0}.\frac{\alpha}{s_c}.\left(\frac{R-s_0}{s_c}\right)^{\alpha-1}.e^{\left[-\frac{V}{V_0}.\left(\frac{R-s_0}{s_c}\right)^{\alpha}\right]} \tag{2.1.7}$$

We use formula (formula 1.3.14) and obtain: [65-66]:

$$< R > = \int_{s_0}^{\infty} R.P_v(R)dR = s_0 + \int_{s_0}^{\infty} (R - s_0).P_v(R)dR \tag{2.1.8}$$

We substitute with the known formulas from this point and obtain [65-66]:

$$< R > = s_0 + \alpha . \frac{V}{V_0} . \int_{s_0}^{\infty} \left(\frac{R - s_0}{s_c}\right)^{\alpha} . e^{\left[-\frac{V}{V_0} . \left(\frac{R - s_0}{s_c}\right)^{\alpha}\right]} dR \qquad (2.1.9)$$

We perform the laying, according to the formula [65-66]:

$$\beta = \left(\frac{V}{V_0}\right)^{\frac{1}{\alpha}} . \frac{R - s_0}{s_c} \qquad (2.1.10)$$

Then we return to the integral and use the addition [65-66]:

$$< R > = s_0 + s_c . \left(\frac{V_0}{V}\right)^{\frac{1}{\alpha}} . \int_{0}^{\infty} \alpha . \beta^{\alpha} . e^{-\beta^{\alpha}} d\beta \qquad (2.1.11)$$

After appropriate transformations, we finally obtain [65-66]:

$$< R > = s_0 + s_c . \left(\frac{V_0}{V}\right)^{\frac{1}{\alpha}} . \Gamma\left(1 + \frac{1}{\alpha}\right) \qquad (2.1.12)$$

where: **Γ(x)** is a mathematical function known as the "gamma function" [69-72].

If we continue with mathematical reasoning, we also obtain the mean square by the formula (using formula 1.3.5 and formula 2.1.6) [65-66]:

$$\sigma_R = s_c . \left(\frac{V_0}{V}\right)^{\frac{1}{\alpha}} . \sqrt{\Gamma . \left(1 + \frac{2}{\alpha}\right) - \Gamma^2 . \left(1 - \frac{1}{\alpha}\right)} \qquad (2.1.13)$$

Now that we have the formulas available, we establish the existence of the theoretical analogue of the coefficient of variation, which is:

$$w_R = \frac{\sigma_R}{< R >} \qquad (2.1.14)$$

This ratio represents the standard itself (according to mathematics) and most accurately reflects the scattering of the destruction limit in bodies of equal volume.

2. Principle of indestructibility

To determine the strength of a material (in this case steel), a sufficient number of steel elements are tested according to a relevant standard. To determine the load on a given element, relevant operational loads of various types of loads and combinations are carried out. This set determines the distribution function for the strength of steel under the relevant type of load and determines the probability estimate for the occurrence of a given steel element. Destruction is assumed to occur in a dangerous state of steel, which can be ductile (tough) or brittle.

The principle for obtaining an estimate of what type the destruction will be or a dangerous state will occur was proposed by N. S. Streletsky [60, 65-66]. The proposed principle is fully consistent with the principles of statistical mathematics and mechanics. We know the distribution curves of the probability density **p(R)** of the destruction strength and of the load **p(S)**. The hypothesis is introduced that the two curves (fig.2.2.1) have zero ordinates for negative values (entirely in the positive values of the abscissa). There is a common point between the two curves $\mathbf{S_0 = R_0}$ – the probability of obtaining a failure limit less than $\mathbf{R_0}$ is according to the formula [60, 65-66]:

$$w_1 = \int_0^{R_0} p(R)dR \tag{2.2.1}$$

Similarly, the probability of a load greater than S_0 is given by the formula [60, 65-66]:

$$w_2 = \int_{S_0}^{\infty} p(S)dS \tag{2.2.2}$$

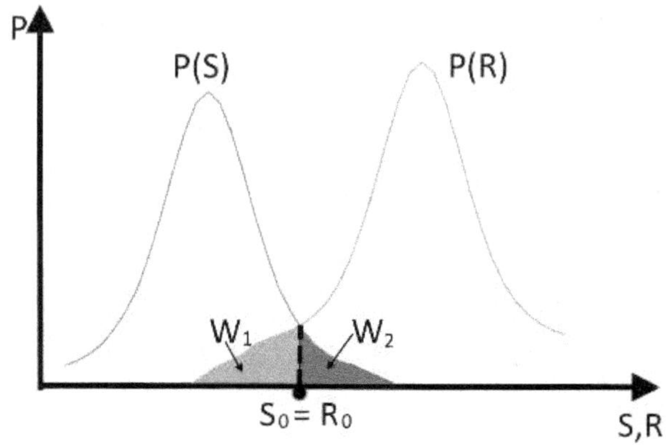

Figure 2.2.1 Distribution curves for the probability density of the load **(P(S))** and the failure limit **(P(R))**

Using the probability multiplication theorem, the value of w1.w2 gives the probability of the simultaneous occurrence of the two events – $\mathbf{S > S_0}$ and $\mathbf{R < R_0}$, and it should be noted that each of them will cause the destruction of the element. The destruction of the element can also occur at $\mathbf{S > R}$ under the following conditions $\mathbf{S < S_0}$ and $\mathbf{R > R_0}$. From this we can conclude that the probability **P(-)** for the occurrence of destruction necessarily satisfies the condition: [60, 65-66]:

$$P(-) > w_1.w_2 \tag{2.2.3}$$

Using probability theory, the product **(1-w_1).(1-w_2)** calculates the probability of the event **S < S_0** and **R > R_0** occurring, i.e. this is the probability that the destruction **P(+)** will not occur:

$$P(+) > (1 - w_1).(1 - w_2) \qquad (2.2.4)$$

Combining the two conditions, we get:

$$w_1.w_2 < P(-) < w_1 + w_2 - w_1.w_2 \qquad (2.2.5)$$

The greatness **Θ = 1 - w_1.w_2** is named by Streletsky (the author) as a guarantee of indestructibility [60, 65-66].

3. Principle of exceeding the strength value

In the exploitation condition of the steel element, there is some excess of the expected stresses in a stationary random process. Even if more than 100,000 elements are tested in tension, each of them will show a different value of the strength for which it was intended. The excess of the catalog-specified strength, which according to the regulatory code requirements is with a 95% probability, as a type of mathematical problem was solved by Rolf-Dieter Reiss and Michael Thomas [64].

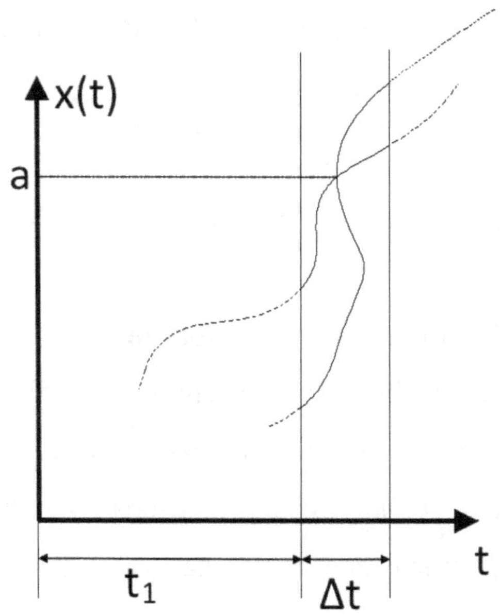

Figure 2.3.1 Distribution curves for the probability density of the load **(P(S))** and the failure limits **(P(R))**

The principle is based on the fact that **x(t)** is a stationary random process and then $\mathbf{v_a}$ is the expected number of crossings with a positive slope at the level **x=a**. There is some small-time interval **Δt** in which this process makes crossings with a positive slope (fig.2.3.1).

Initially, **Δt** is a sufficiently small interval satisfying the condition, we can realize each interval with a straight line without accumulating a sensitive error (under a sensitive error is an error over 5%). Thus, a given realization of the process, if it crosses the line **$t=t_1$** below the level **x=a**, then it will cross it exactly in the interval **Δt** with a positive slope and will fulfill the condition that the slope $\lambda > \frac{a-x}{\Delta t}$, from here it is established to what extent the condition will be fulfilled. In order to realize this condition, the favorable combination between **x** and **λ** should be established. These are random variables, at a fixed value of **t**. The probability (according to mathematics) that **x** is in the interval **x, x+dx** and at the same time **λ.** to be in an interval **λ, λ+dλ** is the function **p(x, λ)dxdλ.**, which (see chapter I) **p(x, λ)** is a joint two-dimensional probability density. But for the above condition to be fulfilled, there must be a favorable combination between **x** and **λ.** Continuing with the reasoning, the expected number of intersections with a positive slope at level **x=a**, at **Δt** is according to the formula [64-66]:

$$\int_{0}^{\infty} d\lambda . \int_{a-\lambda.\Delta t}^{a} p(x,\lambda)dx \qquad (2.3.1)$$

To find the expected number of crossings per unit time in va, we should divide the above expression by **Δt** and make a boundary transition **Δt – 0** and obtain the formula [64-66]:

$$v_a = \int_{0}^{\infty} \lambda . p(a,\lambda)d\lambda \qquad (2.3.2)$$

Formula 2.3.2 is known as Reiss' formula and is practically applicable in all spheres of human activity.

It should be noted that the expected number of exceedances is the basis for determining the favorable realization of the event. In reality, ensemble averaging is established at a precisely defined point in time. It is specified that the stationary process v_a does not depend on time **t** – the average number of exceedances is a constant value, but this does not mean that the average number of exceedances per unit time obtained from an arbitrarily chosen realization by averaging over time would be equal to v_a. If this happens, it means that the process is ergodic.

4. Principle of destruction as a result of a random process

Safety or reliability refers to the ability of a mechanical, electrical, electronic or other system to perform a specific function under specific operating conditions. The cessation of even one of the intended functions is called a malfunction. The term "reliability" entered the Bulgarian technical literature about 40 years ago. This is because the word "reliability" is associated with "hope" or "uncertain expectations".

The Russian term "надежность", officially translated into Bulgarian, means something different and reliable. The English word 'reliability' also contains such nuances. Therefore, it is better to use the term (word) "safety", although the term "reliability" has gained some popularity [66]. The reliability theory is closely related to probability theory. It is an indisputable fact that both the external operating conditions (including all available influences) and the parameters of the target system are stochastic in the way they interact and are created. Therefore, failures are considered as random events, and the probability of failure is a quantitative measure of safety, also called reliability. Closely related to this is the concept of durability, the period of time during which a system can operate "without failure", i.e. the time from the start of operation to "failure" [53-66].

In steel structures, damage can lead to failure. Structures often fail as a result of a constant accumulation of damage, such as residual deformations, reduction of active parts due to wear, and the initiation and propagation of cracks. Such accumulation of damage is a random process and as a result the probability of failure-free operation decreases significantly after a certain period of time. This in itself is not threatening, as long as this probability (the safety of the structure) does not fall below a certain regulatory requirements normative value, which must be adequately justified.

Therefore, the failure of a structure under the influence of random loads is not the only possibility for its failure, nor even the most common method, but it is still one of the possible ways. The collapse can also be reactivated by short-term non-stationary random processes, such as seismic or explosive loading. These can be very short-lived and completely render the term "damage accumulation" meaningless [65-66].

As a first approach, it is justified to look for the probability of exceeding (even once) some dangerous level, at which failure occurs [65-66].

The second approach to the failure of the structure consists in the accumulation of residual deformations under the action of stationary or near-stationary processes [65].

The third approach to structural failure is very common. This is fatigue failure under steady-state processes or under loading close to steady-state. Here again, there is the problem of damage accumulation, and at a relatively low level, often not even exceeding the elastic limit. The "failure" occurs as a result of the characteristic microscopic propagation of cracks [65-66].

The probability of overloading above the specified permissible is $\mathbf{Q\ (S > a|t)}$, since in an interval of $\mathbf{t}$ (time) the load function $\mathbf{S(t)}$ exceeds the permissible load which is determined by the level a (maximum permissible load). I apply Reiss' formula (for.2.3.2) for the average number of exceedances in $\mathbf{v_a}$ per unit time at level $\mathbf{a}$, taking into account that Reiss' formula is for a stationary process, but adding the parameter time (to be made dependent on time, as a function) it will acquire the form for a non-stationary process, as follows [65-66]:

$$v_a(t) = \int_0^{\infty} \lambda . p(a, \lambda; t) d\lambda \quad (2.4.1)$$

After using one of the properties of a stationary process, we obtain [65-66]:

$$v_a(t) = p(a) \int_0^{\infty} \lambda . p(\lambda) d\lambda \quad (2.4.2)$$

To find the average number of exceedances **n(a|t)** for a given time interval **t** at the corresponding level **a**. Since the process has already become non-stationary i.e. $\mathbf{v_a = v_a(t)}$ and for this purpose we need an interval **dt** (step) this leads to $\mathbf{v_a(t)dt}$. According to mathematics we should integrate:

$$n(a|t) = \int_0^{t} dt . \int_0^{\infty} \lambda . p(a, \lambda; t) d\lambda \quad (2.4.3)$$

From the equation (eq.2.4.3) it follows that **n(a|t)** is a decreasing function with respect to a. If we assume that $a_0(t)$ is the place (level) at which the condition that the excesses are equal to unity is fulfilled, $a_0(t)$ is a root of the equation, i.e.

$$n(a|t) = 1 \quad (2.4.4)$$

In this case, there will be two possibilities. The first is $\mathbf{a \leq a_0}$ which means that **n(a|t) => 1** – exceeding the permissible load (level) **a** is a certain (reliable) event and in this case **Q (S > a|t) = 1**. In the next possibility we have $\mathbf{a > a_0}$, in this case we assume that gives us the desired probability of **Q (S > a|t)**.

The second approach assumes that there is an excess of the elastic limit of the material, as a result of which the destruction occurs by a random process. The load is above the maximum permissible load for some time and as a result of this circumstance, plastic and residual deformations are formed. For the sake of comprehensibility of the matter, we assume that the following prerequisites are present [65-66]:

* External loads, without exceptions, change proportionally as a general random process in time **S(t)**, which we assume to be a generalized force;

* We will not take into account parasitic (secondary) residual deformations;

* The influence of inertial forces (moments), relaxation and creep are neglected.

Now we formulate a problem: Given the corresponding characteristics of the generalized force, to determine the distribution in a corresponding time interval **(t)** of the residual deformations. The sought distribution is described by a random distribution parameter **δ**, which is called the generalized residual deformation. To simplify the solution, we assume that there is a deterministic dependence between **S** and **δ**, which dependence is also known to us under monotonic loading, i.e. [65-66]:

$$S = \psi(\delta) \text{ и } \delta = \varphi(S) \quad (2.4.5)$$

Let's assume that **a(t)** is the maximum value of **S(t)** in the interval **(0; t)** for the process, but it necessarily includes unloading. The equation [65-66] is obtained:

$$a(t) = \psi[\delta(t)] \text{ и } \delta(t) = \varphi[a(t)] \quad (2.4.6)$$

It is known from practice that the initial moment is the moment at which **δ = 0**. Given the above formula (form.2.4.6) and the formulas from probability theory, we find the probability density of the generalized deformation [65-66]:

$$p[\delta(t)] = p\{\psi[\delta(t)]\}.\left|\frac{d\psi(\delta)}{d\delta}\right| \quad (2.4.7)$$

Based on formula (formula 2.4.7), we can also find the mathematical expectation of the generalized deformation δ at the end of the time interval (0;t), namely [65-66]:

$$< \delta(t) > = \int_{0}^{\infty} \delta.p(\delta(t)d\delta \quad (2.4.8)$$

This formula is confirmed, since it can also be obtained by integrating the variable, and then the mathematical expectation for **<φ(b)>** of the existing function **φ(b)**, of a random variable **b**.

$$< \varphi(b) > = \int_{-\infty}^{+\infty} \varphi(b).p(b)db \quad (2.4.9)$$

or written another way [65-66]:

$$< \delta(t) > = \int_{b_0(t)}^{\infty} \varphi(b).p[b(t)]db \quad (2.4.10)$$

The lower limit of the integral $b_0(t)$ has a value equal to zero, which corresponds to a residual deformation equal to zero, i.e. a load where there are no deformations.

The third case, namely fatigue failure due to the presence of microcrack propagation in the element, is the most difficult. The S-N curve, also known as the steel fatigue curve, has been known for more than 75 years, and it defines as a graphical relationship the required number of cycles N and the required stress level S for the failure stress to occur. It is a hyperbolic curve and is analytically written in the form [65-66]:

$$N.S^b = c \tag{2.4.11}$$

where: **b** and **c** are constants depending on the type of steel, and **S** is the amplitude of the stresses. In practice, it is more convenient to write it in logarithmic form:

$$\log N + b.\log S = \log c \tag{2.4.12}$$

In this curve recording, it is a linear function, which corresponds to the deterministic approach. In random processes, we cannot have a single S-N curve, because there are many such curves available, which actually exclude the possibility of using this method. It is correct to assume that there is a statistical distribution of the required number of cycles N corresponding to the given stress level. The main problem is to provide a suitable way to determine the moment of fatigue failure. To date, no single solution is known for this case. Relatively satisfactory results are given by the hypothesis of Palmgren–Miner (also known as Miner's rule or the linear damage rule) [65-66]. According to it, we have a certain number of cycles n and a corresponding amplitude S (of stress) in the steel, there are a certain number of damages (which are cumulative) that begin to take away from the existence of the steel, which is equal to $\frac{n}{N}$, where **N** is the maximum number of cycles at constant amplitude **S**, which the steel can absorb according to the curve **S-N**. Also, the specified number of failures contains other failures of a different nature. Mathematically, it is written as follows [65-66]:

$$D = \sum_{i=1}^{m} \frac{n_i}{N_i} \tag{2.4.13}$$

where: **D** – general damage; $\mathbf{n_i}$ – number of cycles; $\mathbf{N_i}$ – destructive number of cycles

According to the Palmgren–Miner hypothesis, when **D** becomes equal to unity **(D=1)** then fatigue failure occurs. With this definition, the order in which the different stress levels are applied does not matter, which allows us to use it for a random process [65-66].

For an experimental specimen (tube), its stresses represent a narrow-spectrum random stationary process. There is a mathematical expectation **<D(t)>** of the total failure in a corresponding time range **(t)** [65].

The expected real frequency of the process (established per unit time in number of cycles) v_a (the expected number of crossings of the abscissa axis with a positive slope per unit time) for the corresponding time **t** we have $\mathbf{t.v_a}$ number of crossings. The probability density of the maxima **p(a)**, which is one for each cycle, but is of a different value (size), follows that the expected number of maxima with size a between **a** and **a+da** for one cycle is **p(a).da**, and for $\mathbf{v_a.t}$ cycles is [65-66]:

$$n(a) = v_a.t.p(a)da \tag{2.4.14}$$

Thus, in accordance with the Palmgren–Miner hypothesis – for a single maximum of size **a**, a failure equal to $\frac{1}{N(a)}$, where **N(a)** is the breaking number of cycles at a voltage of a certain amplitude **a**, determined by the curve **S-N**. Then the failure corresponding to the corresponding number of cycles **n(a)** is [65-66]:

$$\frac{n(a)}{N(a)} = v_a.t.\frac{p(a)}{N(a)}da \tag{2.4.15}$$

Where in the mathematical expectation of the total damage D at all possible levels follows the integration, i.e. [65-66]:

$$< D(t) >= v_a.t.\int_0^{\infty}\frac{p(a)}{N(a)}da \tag{2.4.16}$$

The probability density for the maxima (the peaks of the narrow-spectrum random process) is closely related to the estimate of the expected failure. To find this distribution we need to find the difference [65-66]:

$$v_a - v_{a+da} = -\frac{dv_a}{da}da \tag{2.4.17}$$

which gives us the number of intersections of two infinitely close levels **a** and **a+da** respectively and the curve has a positive slope. It is this difference that gives us information about the expected number of peaks between these levels. The expected number of peak values on the ordinate is equal to the expected number of cycles per unit time – $\mathbf{v_a}$. Under this assumption the probability of **p(a)da** for the realization of a peak value (point) between two levels **a** and **a+da** is [65-66]:

$$p(a)da = -\frac{1}{v_a}.\frac{dv_a}{da}da \tag{2.4.18}$$

Applying mathematical statistics and probability theory, we obtain the formula for the distribution of vertices (peaks) of a narrow-spectrum stationary random process with a Gaussian distribution of the zero-mean included, i.e. [65-66]:

$$p(a) = \frac{a}{\sigma_x^2}e^{\left(-\frac{a^2}{2.\sigma_x^2}\right)} \tag{2.4.19}$$

Formula 2.4.19 is the probability density of the Rayleigh distribution (John William Strutt, 3rd Baron Rayleigh) known in probability theory [65].

Returning to the basic formula (formula 2.4.16) and substituting, we obtain [65-66]:

$$N(a) = \frac{c}{a^b} \tag{2.4.20}$$

$$< D(t) > = \frac{v_a}{c.\sigma_x^2}.\int_0^\infty a^{b+1}.e^{\left(-\frac{a^2}{2.\sigma_x^2}\right)}da = \frac{v_a.t}{c}.\left(\sigma_x.\sqrt{2}\right)^b.\Gamma.\left(1+\frac{b}{2}\right) \quad (2.4.21)$$

where: **Γ(x)** is the famous Gamma function (in mathematics).

The above formula leads to the conclusion that the expected total failure is proportional to the number of cycles $\mathbf{t.v_a}$. It depends on the standard $\boldsymbol{\sigma_x}$ for the process and on the cyclic loads characteristic of the material behavior through the constants **b** and **c**. It is also established by the formula that it is possible to determine and calculate the expected durability **d**, the expected interval until fatigue failure occurs, i.e. **<D(d)> = 1** or [65-66]:

$$d = \frac{c}{v_a}.\frac{1}{\left(\sigma_x.\sqrt{2}\right)^b.\Gamma.\left(1+\frac{b}{2}\right)} \quad (2.4.22)$$

Formula 2.4.22, although justified according to mathematics, mechanics and the hypotheses and assumptions known so far, is not ideal and cannot be applied in all cases, i.e. it is of limited use. If the values are high enough, then the formula will give extremely good and accurate results, but if there are low values of the stress amplitude, then the formula will give a fairly significant error (an error of over 10% is significant). This is due to the fact that when accepting the integral in the formula there is a lower limit equal to zero, i.e. negligible stresses and their participation in the damage are not taken into account. In turn, there is a finite **N**, even smaller **S**. In some authors [46-48] they found that if **S < R**, where **R** is the fatigue strength in a symmetric cycle, then the destructive number of cycles is infinitely large (practically non-existent). To avoid this part, it is probably better to make a better approximation of the **S-N** curve, whose analytical form [65] is:

$$N = c_1.\left(\frac{R}{S}\right)^b, \text{при } S \geq R\,;\text{ и } N \rightarrow \infty \text{ при } S \leq R \quad (2.4.23)$$

By performing the substitution (Form.2.4.22), we obtain [65]:

$$d = \frac{c_1.x_0^b}{v_a.\psi(b+2).\,p(x_0^b; b+2)} \quad (2.4.24)$$

where: $\boldsymbol{x_0} = \frac{R}{\sigma_x}$, $\boldsymbol{p(x_0^2, n)}$ is the Peterson distribution function [65], and:

$$\psi(n) = 2^{\frac{n-2}{2}}.\Gamma\left(\frac{n}{2}\right) \quad (2.4.25)$$

Certainly, other improvements of the theory of damage accumulation are known, given in the literature [57]. In one of them, the hypothesis of two stages of destruction is accepted. In the first stage, the conditions for the development of the crack are prepared, and in the second stage the crack develops in depth in the element until its destruction. In accordance with this formulation of the theory, two characteristics of the damage are also introduced – a measure of the structure's non-compaction and a measure of the development of the crack.

5. Principle of polynomial approximation

Undoubtedly, the existence of nonlinear partial differential equations, which are able to describe the development function of a given process (case), is practically established in practice, but the problem comes from an infinite number of degrees of freedom for the respective process. In reality, there are a minimum number of equations (low-order equation systems) or a sufficient number of equations as required by the goals and analysis of a given process. In this part, it should be specified or determined in advance to what degree of approximation the process should extend. Practically in this direction is the existence of geohydrodynamic problems, which are described by nonlinear systems and partial differential equations, their solution is not only by eliminating the spatial variables through spectral decomposition, but also by approximating the sought functions by polynomials to a certain degree. In this way, a complete system of ordinary differential equations is obtained with the coefficients without any approximation. Very often in the practical solution, the so-called the two-dimensional Navier-Stokes equations, which have been known for more than 80 years, but strangely enough have been forgotten and rediscovered many times on different occasions [14-15]. For the sake of understanding the present matter, let us take a process that consists of atmospheric mechanics and geophysical hydrodynamics, and through these equations, numerous mathematical models have been created, which are of the type [6-8,13]:

$$\begin{aligned}
&\frac{\partial U}{\partial t} + U.\frac{\partial U}{\partial x} + V.\frac{\partial U}{\partial y} - f.V = -\frac{\partial H}{\partial x} \\
&\frac{\partial V}{\partial t} + U.\frac{\partial V}{\partial x} + V.\frac{\partial U}{\partial y} + f.V = -\frac{\partial H}{\partial y} \\
&\frac{\partial H}{\partial t} + U.\frac{\partial V}{\partial x} + V.\frac{\partial U}{\partial y} + H.\left(\frac{\partial U}{\partial x} + \frac{\partial V}{\partial y}\right) = 0
\end{aligned} \tag{2.5.1}$$

where: **f=2.ω.sinφ** is the Coriolis parameter [6-8]

Analyzing the equation (eq.2.5.1), it is established that for them there always exists a solution of the form [6-8]:

$$\begin{aligned}
U(x,y,t) &= U_0 + U_1.x + U_2.y \\
V(x,y,t) &= V_0 + V_1.x + V_2.y
\end{aligned} \tag{2.5.2}$$

$$H(x,y,t) = H_{00} + H_{10}.x + H_{20}.y + \frac{1}{2}.H_{20}.x^2 + \frac{1}{2}.H_{22}.y^2 + H_{12}.x.y$$

where: $\mathbf{U_k=U_k(t)}$; $\mathbf{V_k=V_k(t)}$; $\mathbf{H_k=H_k(t)}$; **i,k=1,2**.

In this analysis, we find that the first two equations of eq. 2.5.1 lead to a nonlinear dynamic system of equations as follows [11]:

$$\dot{U_k} + U_k.U_1 + V_k.(U_2 - f) = -H_{1k'}$$
$$\dot{V_k} + V_k.V_2 + U_k.(V_1 + f) = -H_{2k'} \quad (2.5.3)$$

where: к=1,2 and two additional equations

$$\dot{U_0} + U_0.U_1 + V_0.(U_2 - f) = -H_{10'}$$
$$\dot{V_0} + V_0.V_2 + U_0.(V_1 + f) = -H_{20'} \quad (2.5.4)$$

where: $\mathbf{U_{1,2}}$ and $\mathbf{V_{1,2}}$ are known from the previous systematic solution of the process.

$\mathbf{U_0}$ and $\mathbf{V_0}$ by their very nature they represent translational motion (simple), therefore we assume $\mathbf{H_{10}=H_{20}=0}$ и $\mathbf{U_0=V_0=0}$, therefore, only equations 2.5.3 remain. To avoid confusion, it has new variables, i.e. [1-5]:

$$D(t) = \frac{\partial U}{\partial x} + \frac{\partial V}{\partial y} = U_1 + V_2\dot{V_0} + V_0.V_2 + U_0.(V_1 + f) = -H_{20'} \quad (2.5.5)$$

$$\zeta(t) = \frac{\partial V}{\partial x} - \frac{\partial U}{\partial y} = V_1 + U_2 \quad (2.5.6)$$

$$F(t) = \frac{\partial U}{\partial x} - \frac{\partial V}{\partial y} = U_1 - V_2 \quad (2.5.7)$$

$$M(t) = \frac{\partial V}{\partial x} + \frac{\partial U}{\partial y} = V_1 + U_2 \quad (2.5.8)$$

The first two equations (eq.2.5.5 and eq.2.5.6) are essentially divergence, and the last two are essentially deformation. By making a setting of the form:

$\mathbf{S(t)=H_{11}+H_{22}}$; $\mathbf{R(t)=H_{11}+H_{22}}$; $\mathbf{Q(t)=2.H_{12}}$ – finally [1-5,9-10]:

$$\dot{\zeta} = -D(f + \zeta) \quad (2.5.9)$$

$$\dot{F} = -D.F + f.M - R \quad (2.5.10)$$

$$\dot{M} = -D.M + f.F - Q \quad (2.5.11)$$

$$\dot{D} = f.\zeta - S - \frac{1}{2}(D^2 + F^2 + M^2 - \zeta^2) \quad (2.5.12)$$

Hypothetically, the equations (equations 2.5.9, 2.5.10, 2.5.11 and 2.5.12) are a coupled nonlinear dynamic system (possibly the most complex variant), then the only physically meaningful steady-state solution is $\mathbf{(Q = 0)}$ [1-5,9-10]:

$$\overline{D} = \overline{F} = 0 \; ; \; \overline{M} = \frac{R}{f} \; ; \; \overline{\zeta} = -f + \sqrt{f^2 + 2.S + \frac{R^2}{f^2}} \qquad (2.5.13)$$

This solution always exists under the condition **S > 0**, i.e. the pressure must have a positive sign. In cases of negative pressure, we use the mathematical model of the modular solution, i.e. **S= -|S| < 0**, for **R** and **S** - some restrictions should be imposed, namely: $0 \leq \mathbf{2}.|S| - \frac{R^2}{f^2} \leq f^2$ so that he can $\overline{\zeta}$ to be a real quantity and the resulting solution to be sustainable.

The complete system of the process is in equation 2.5.13, from here it follows that four more equations should be added to the equation, as follows [1-5,9-10]:

$$\dot{S} = -2.D.S - F.R - M.Q \qquad (2.5.14)$$

$$\dot{R} = -2.D.R - F.S - \zeta.Q \qquad (2.5.15)$$

$$\dot{Q} = -2.D.Q - M.S - \zeta.R \qquad (2.5.16)$$

$$\dot{H_w} = -D(t).H_{00} \qquad (2.5.17)$$

This system that is finally obtained is a system of 8 nonlinear differential equations and its solution is not easy, even impossible for a practicing engineer. The peculiarity of the differential equations themselves admits some invariants, namely [1-5,9-10]:

$$\frac{d\Delta}{dt} = -4.D(t).\Delta \qquad (2.5.18)$$

where: $\mathbf{\Delta t = H_{11}.H_{22} - H_{12}^2 = \frac{1}{4}.(S^2 - R^2 - Q^2)}$ – this is a characteristic of a plane curve of the second degree. Combining eq.2.5.18 and eq.2.5.17 with the first equations (Eq.2.5.9, 2.5.10, 2.5.11 and 2.5.12), we obtain:

$$\frac{\Delta(t)}{H_{00}^4(t)} = \text{const} \; ; \; \frac{\Delta(\zeta)}{\zeta^4} = \text{const} \qquad (2.5.19)$$

where: the constants are determined by the initial conditions.

More than 75 years ago, the Russian mathematician and meteorologist Ilya A. Kibel laid the foundations of numerical methods in the field of meteorology [6-8] by proposing a system of partial nonlinear differential equations of the form:

$$\frac{\partial T}{\partial t} = \frac{\partial T}{\partial x}.\frac{\partial \theta}{\partial y} - \frac{\partial T}{\partial y}.\frac{\partial \theta}{\partial x} = J_{xy}(T, \theta) \qquad (2.5.20)$$

$$\frac{\partial \theta}{\partial t} = \frac{\partial T}{\partial x}.\frac{\partial T}{\partial y}.\left(\frac{\partial^2 \theta}{\partial y^2} - \frac{\partial^2 \theta}{\partial x^2}\right) + \left[\left(\frac{\partial T}{\partial x}\right)^2 - \left(\frac{\partial T}{\partial y}\right)^2\right].\frac{\partial^2 \theta}{\partial x \, \partial y} \qquad (2.5.21)$$

The functions **T** and **θ** are represented as second degree polynomials in x,y analogously to those in equation 2.5.2. For **H(x,y,t)** then these partial nonlinear differential equations of Kibel automatically form a system of 12 ordinary differential equations, half of which necessarily form a closed system for an exact analytical solution, assuming **(T, *θ*) = A**, it is obtained:

$$A(x, y, t) = A_0 + \frac{1}{2}.A_1.x^2 + \frac{1}{2}.A_2.y^2 + A_3.x.y \tag{2.5.22}$$

Finite difference approximation is also used to solve nonlinear dynamical systems. One such process is the logistic equation, of the form:

$$X = r.\dot{X}(1 - X) \tag{2.5.23}$$

Its exact solution is known and it is:

$$X(t) = \frac{C.e^{r.t}}{1+C.e^{r.t}} \; ; C = \frac{X_0}{1-X_0} \; ; \; X_0 = X(0) \tag{2.5.24}$$

In [11] it is stated that the discrete analogue has a completely different property. At **r = 1**, and the first derivative is approximated by central difference according to equation [11]:

$$\frac{X_{n+1} - X_{n-1}}{2.\Delta t} = X_n.(1 - X_n) \tag{2.5.25}$$

We write the result thus obtained in a two-dimensional discrete nonlinear dynamical system of the form [11]:

$$X_{n+1} = 2.\Delta t.(1 - X_n).X_n + Y_n \; ; \; Y_{n+1} = X_n \tag{2.5.26}$$

In this equation there are two stationary points **(0,0)** and **(1,1)**, these points are hyperbolic (unstable). The difference **X(t).X$_n$** does not tend to 1, and [11] for each **Δt** the system has chaotic behavior. For differential processes there is a great variety, this gives us reason to use the properties of the discrete analogue of Eq.2.5.23 and so the dependence is on the type of process used. Using what was published in [11] instead of Eq.2.5.25, we obtain:

$$(1 - \mu).\frac{X_{n+1} - X_{n-1}}{2.\Delta t} + \mu.\frac{X_{n+1} - X_n}{\Delta t} = X_n.(1 - X_n) \tag{2.5.27}$$

At **μ = 0** we return to equation 2.5.25, but at **μ = 1** is the Euler scheme (also called the forward Euler method) – equation 2.5.27 coincides with the logistic equation and there is a solution depending on the value of **Δt.** In all other cases, the scheme is of mixed type and is equivalent to the system:

$$X_{n+1} = \frac{1-\mu}{1+\mu}.Y_n + \frac{2}{1+\mu}.[(\mu + \Delta t).X_n - \Delta t.X_n^2] \; ; \; Y_{n+1} = X_n \tag{2.5.28}$$

The points **(0,0)** and **(1,1)**, which are stationary, the existence of an asymptotically stable system is when the conditions **0 < Δt < 1** and **μ > 0,5.Δt** are met. The solution of Eq.2.5.27 is performed for arbitrarily large values of **n**. The chaotic solution is at **μ < 0,5.Δt**, which is explained by the help of the affine coordinate transformation by the equation [103, 110]:

$$(X_n, Y_n) = \frac{\mu+\Delta t}{2.\Delta t}.\left(\frac{\Delta t-\mu}{1+\mu}.U_n + 1, \frac{\Delta t-\mu}{1-\mu}.V_n + 1\right) \tag{2.5.29}$$

By this inversion, one automatically transforms eq. 2.5.29 into the familiar Henon function [103, 110] (also known as the "Henon map" or "Hénon–Pomeau attractor/map" and applies to a discrete-time dynamical system), namely:

$$U_{n+1} = 1 + V_n - a.U_n^2 \ ; \ V_{n+1} = b.U_n \tag{2.5.30}$$

where:

$$a = \frac{(\Delta t)^2 - \mu^2}{(1+\mu)^2} \ ; \ b = \frac{1-\mu}{1+\mu}$$

The approach considered so far shows a perfect way to solve nonlinear dynamic processes by a nonlinear numerical method for physical or other effects. Certainly, numerical approximation equations are an elegant way, especially when we are looking for a solution in a given time interval **0 < t < T**, through which, with a sufficiently small step **Δt**, they compose with sufficient accuracy a reliable function of the development of the process in time.

The existence of computational instability of the solution is overcome by choosing the step **Δt** depending on the respective case (a smaller step leads to closer data). The only danger that exists is to fall into a situation when time tends to infinity and then the so-called computational chaos (uncertainty) occurs. In order to avoid computational chaos, several construction rules should be observed.

The mathematical model of a specific phenomenon or process, described by one or more equations, directly compiled on the basis of test data, is converted into a polynomial function, observing the principle that when connecting the data points there should be without "sharp corners" (fig.2.5.1), i.e. the curve should run smoothly, since each process, compiled on the basis of this type of equations, is continuous [14-15].

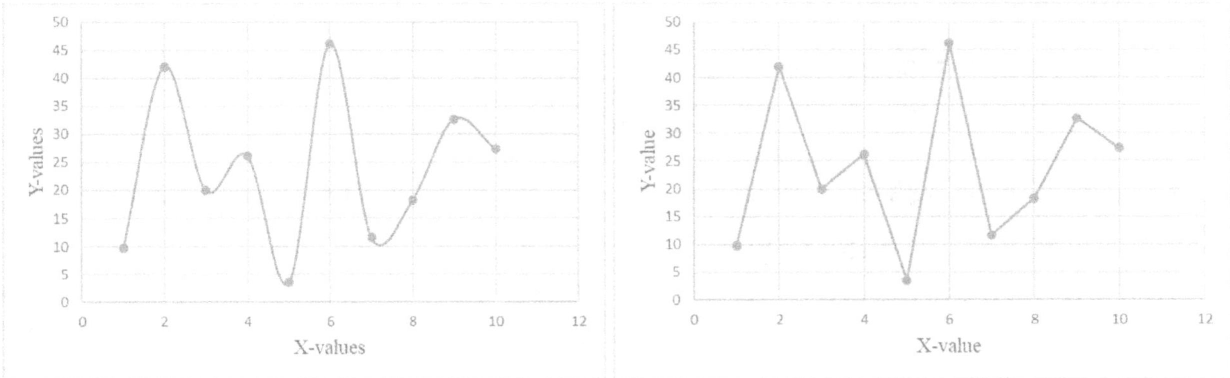

(a) (b)

Figure 2.5.1 a) polynomial function without "sharp corners" (continuous); b) polynomial function with "sharp corners" (no continuous)

6. Principle of transition from a deterministic random system to linear equations

One of the most common problems in statistical mechanics is the study of the response of stationary linear systems to a stationary random effect. The methods described in the previous paragraphs are fully applicable to this problem. It is interesting to obtain simple relations that give a solution to the problem in closed form. Let us recall the equation of the stationary system (eq.1.2.5) [65]:

$$L.u = q \tag{2.6.1}$$

We denote **q(t)** a stationary random process. Since the response of a linear system to the mathematical expectation of the input process is determined separately from the equation, then without loss of generality we can assume that the process is centered, i.e. **<q(t)>= 0**. In this case, we represent the process in the form of a stochastic Fourier integral, i.e.:

$$q(t) = \int_{-\infty}^{+\infty} Q(w).e^{i.w.t}\,dw \tag{2.6.2}$$

And it should satisfy the condition [65]:

$$< Q^*(w).Q(w') > = S_q(w).\delta(w - w') \tag{2.6.3}$$

$$K_q(t_1,t_2) = \int_{-\infty}^{+\infty}\int_{-\infty}^{+\infty} < Q^*(w).Q(w') >.e^{i.(t_1 - w.t_2)}dwdw' \tag{2.6.4}$$

$$S_q(w) = \frac{1}{2.\pi}.\int_{-\infty}^{+\infty} K_q(\tau).e^{-l.w.\tau d\tau} \tag{2.6.5}$$

Hypothetically, we assume that the operator **L** is an operator with constant coefficients and is of polynomial type, i.e. of the form [14-15,65]

$$L = a_0.\frac{d^v}{dt^v} + a_1.\frac{d^{v-1}}{dt^{v-1}} + a_2.\frac{d^{v-2}}{dt^{v-2}} + \cdots..+a_{v-1}.\frac{d}{dt} + a_v \tag{2.6.6}$$

$$L.\psi = e^{i.w.t} \tag{2.6.7}$$

The introduced hypothesis corresponds to a function that is in the limits +∞ and - ∞, then the solution of this function is [14-15,65-66]:

$$\psi(t|w) = \frac{e^{i.w.t}}{L(iw)} \tag{2.6.8}$$

By applying differential mathematics, the operator L becomes a differentiation operator, i.e. [14-15,65-66]:

$$L(iw) = a_0(iw)^v + a_1(iw)^{v-1} + a_2(iw)^{v-2} + \cdots \ldots\ldots\ldots + a_{v-1}(iw) + a_0 \tag{2.6.9}$$

$$K_u(\tau) = \int_{-\infty}^{+\infty} \frac{S_q(w).\, e^{i.w.\tau} dw}{|L(iw)|^2} \tag{2.6.10}$$

By comparing it with the result, expressed in the correlation function of the output process, through its spectral density [14-15,65-66]:

$$K_u(\tau) = \int_{-\infty}^{+\infty} S_u(w).\, e^{i.w.\tau} dw \tag{2.6.11}$$

Then the relationship between the input-output processes and the spectral density is determined by the formula [14-15,65-66]:

$$S_u(w) = \frac{S_q(w)}{|L(iw)|^2} \tag{2.6.12}$$

Formula 2.6.12 is the basic relationship that provides a solution to the problem of the transition of a stationary process through a stationary polynomial system. Once the spectral density of the output process is found, its correlation and moment functions, the correlation functions of the derivatives of the output process, and the variations of the output process and its derivatives are easily calculated.

7. Principle for solving nonlinear problems using a stochastic approach

The solution of nonlinear problems for distributed systems is associated with considerable difficulties. Only very few problems have an exact closed-form solution. Therefore, approximate methods prevail in this area. Most of the approximate methods for statistical mechanics of discrete systems, which were mentioned and described above, can be extended to distributed systems. Such is, for example, the method of closing a chain of equations with respect to moment functions, using statistical hypotheses about the relationship between the highest moment functions. The method of small parameters, the method of statistical linearization and some others are also easily generalized to problems that are described by partial differential equations [65-112].

In nonlinear stochastic problems of distributed systems, a solution is sought that allows the solution to be in closed form. For an infinitely long circular cylindrical shell that is loaded with an axisymmetric random load with intensity that is a homogeneous random ergodic function of the axial coordinate x. We assume that the attachment of the shell is such that over a sufficiently large length the mutual axial displacements of the points of the mid-surface are on average equal to zero. The decisive nonlinear factor will be the axial force preventing the clustering of the section [65]:

$$N = \lim_{a\to\infty} \frac{E.h}{2.a} \int_{-\frac{a}{2}}^{\frac{a}{2}} \left(\frac{dw}{dx}\right)^2 dx \tag{2.7.1}$$

$$\lim_{a\to\infty}\frac{1}{a}\int_{-\frac{a}{2}}^{\frac{a}{2}}\left(\frac{dw}{dx}\right)^2 =< v^2 >= \sigma_v^2 \quad (2.7.2)$$

The equation of axisymmetric deformation of elements, taking into account the axial force, has the equation [65]

$$D.\frac{d^4w}{dx^4}+\frac{E.h}{R^2}.w-N.\frac{d^2w}{dx^2}=q \quad (2.7.3)$$

No additional conditions are imposed on the function, except for the requirement that it be bounded at +∞.

Equation 2.7.3 with the parameter N defined according to (eq.2.7.1), is nonlinear, since **N** is expressed nonlinearly by **σ(x)**. However, the nonlinearity here is of a special nature and does not significantly complicate the solution of the problem compared to the polynomial problem. Spectral densities for the functions **q(x)** and **w(x)** (fig. 1.2.2) [65]:

$$S_q(\chi)=\frac{1}{2.\pi}.\int_{-\infty}^{+\infty}K_q(\xi).e^{-i.\chi.\xi}\,d\xi \quad (2.7.4)$$

$$S_w(\chi)=\frac{1}{2.\pi}.\int_{-\infty}^{+\infty}K_w(\xi).e^{-i.\chi.\xi}\,d\xi \quad (2.7.5)$$

These densities are related to the equation following from Eq. 2.7.3. The solution of this equation gives [65]:

$$S_w(\chi)=\frac{S_q(\chi)}{\left(\chi^4.D+\frac{E.h}{R^2}+\frac{1}{2}\chi^2.E.h.\sigma_v^2\right)^2} \quad (2.7.6)$$

The right side of formula 2.7.6 contains the parameter a, which according to equation 2.7.5 is expressed by the spectral density

$$\sigma_v^2=\int_{-\infty}^{+\infty}\chi^2.S_w(\chi).\,d\chi \quad (2.7.7)$$

Substituting here into eq.2.7.6, we arrive at an equation for the parameter $\boldsymbol{\sigma_v^2}$ [65]:

$$\sigma_v^2=\int_{-\infty}^{+\infty}\frac{\chi^2.S_w(\chi).d\chi}{\left(\chi^4.D+\frac{E.h}{R^2}+\frac{1}{2}\chi^2.E.h.\sigma_v^2\right)^2} \quad (2.7.8)$$

In simple cases, the integral included in this equation is calculated in this way [65]:

$$\sigma_v^2=\frac{c}{(1+\chi.\sigma_v^2)^{\frac{3}{2}}} \quad (2.7.9)$$

where:

$$c = \frac{\pi.s}{2^{\frac{5}{2}}.D^2.\chi_0^5} \quad и \ \chi = \frac{1}{4}.\chi_0^2.R^2 \qquad \chi_0 = \left(\frac{E.h}{R^2.D}\right)^{\frac{1}{4}}$$

The problem considered above turned out to be so accessible to solution not only because the desired function depends on only one variable **x**. It is easy to see that the original equation 2.7.3 is by its nature statistically linearized. The nonlinearity enters it through the mean square of the derivative **ω** and for a given mean square the equation is linear with respect to the desired function **ω(x)**.

This theoretical solution serves as a standard for comparing different approximate methods. For example, the small parameter method - the intensity of the external load is sufficiently small and we assume that **q=μ.q(x)**, where **μ** is a small parameter, we should look for a solution to equation 2.7.3 in the form of a series of equations [65-92] of the type:

$$w = \mu.w_1(x) + \mu^2.w_2(x) + \mu^3.w_3(x) + \cdots \ldots\ldots\ldots\ldots\ldots + \mu^n.w_n(x) \tag{2.7.10}$$

$$D.\frac{d^4w_1}{dx^4} + \frac{E.h}{R^2}.w_1 = q_1 \tag{2.7.11}$$

$$D.\frac{d^4.w_3}{dx^2} + \frac{E.h}{R^2}.w_3 = \frac{1}{2}.E.h.a_{11}.\frac{d^2w_1}{dx^2} \tag{2.7.12}$$

$$D.\frac{d^4.w_5}{dx^4} + \frac{E.h.w_5}{R^2} = \frac{1}{2}.E.h.\left(a_{11}.\frac{d^2.w_3}{dx^2} + 2.a_{13}\frac{d^2w_1}{dx^2}\right) \tag{2.7.13}$$

$$a_{jk} = \langle\frac{dw_j}{dx}.\frac{dw_k}{dx}\rangle \tag{2.7.14}$$

One of the most effective approximate methods is the method of reduction to discrete systems using expansions through equations of the type of equations 2.6.9, representing the solution of a nonlinear problem in the form of a series of polynomial sections through equation 2.6.9 and applying one of the variational methods, a system of ordinary differential or algebraic equations in terms of generalized coordinates is reached. Additional studies are carried out using methods of statistical mechanics of discrete systems.

8. Summary

From the briefly mentioned basic principles, it is established that there are mathematical models through which, based on experimental data and the corresponding processing, they provide with sufficient computational accuracy for practice the possibilities of modeling a variety of processes that exist in practice. The use of these approaches in combination with statistical mechanics leads to the establishment of corresponding trends through functional curves (dependencies) in a dependent variable from the temporal impact of the respective influence.

CHAPTER III. POLYNOMIAL FUNCTIONS

To establish the polynomial equations or functions concerning the change in the stress-strain diagrams of the material and the Young's modulus (E-modulus, modulus of elasticity,modulus of linear deformation), experimental data (given in Appendix No. 1) are used, from which the values of strength and relative deformation are taken. They are processed by applying the stochastic method and the average method, and the E-modulus is calculated according to the formulas, using the values obtained by the stochastic method and the average method. The final result is presented in a tables.

The diagrammatic dependence is drawn up and, using the polynomial approximation, the equation of the functions is established, according to which function the corresponding parameter will change depending on the corrosion influence and category over time.

1. Corrosion category C1

The corrosion category does not occur in natural climatic conditions, but is intended for indoor climatic conditions in buildings according to the ISO 12944-5 standard, such as heated production buildings, offices, shops, schools, hotels, etc.

1.1. Point of the proportional limit

On table 3.1.1 is present a result after processing by stochastic method and average method. In fig. 3.1.1(a) shows the relationship between the change in relative strain versus the action time of the corrosion category, and in fig. 3.1.1(b) shows the relationship between the change in strength over time due to the corrosion effect of category.

Table 3.1.1 Result after processing [119]

	Stochastic results									
time, [months]	0	9138	14769	24923	34892	46154	54000	64615	73108	77262
ε, [mm]	2.070	2.521	2.727	2.595	2.151	2.870	2.432	2.296	1.414	1.331
σ, [MPa]	218.90	211.90	209.30	193.00	224.00	218.00	213.40	234.90	173.90	205.90
	Average results									
time, [months]	0	9138	14769	24923	34892	46154	54000	64615	73108	77262
ε, [mm]	2.130	2.543	2.619	2.664	2.039	2.511	2.423	1.907	1.299	1.282
σ, [MPa]	205.70	211.30	208.40	191.30	212.40	206.40	220.50	200.20	156.90	176.60

Using the polynomial approximation [14-15,114-119] from fig. 3.1.1, functional dependence for the strain and stress point in dependence of the time of the influence of the corrosion (in months) is established in functions or equations.

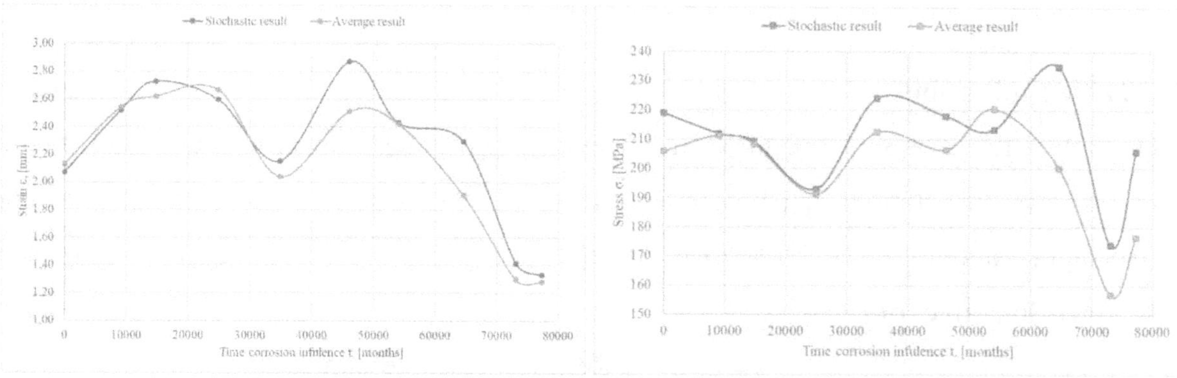

(a) (b)

Figure 3.1.1 (a) Graph of the variation of the strain deformation depending on the corrosion effect over time [119]; (b) Graph of the change of stress strength depending on the corrosion effect over time [119];

1.1.1. Stochastic functions

- Strain function [119]

$$\begin{aligned}\varepsilon(t) = {} & 4.75622394850429.10^{-40}.t^9 - 1.677658016573874.10^{-34}.t^8 + 2.47279097532.10^{-29}.t^7 \\ & - 1.9733039162081.10^{-24}.t^6 + 9.231159634032798.10^{-20}.t^5 - 2.5607472917642749.10^{-15}.t^4 \\ & + 4.0599624820046138.10^{-11}.t^3 - 3.34033365812319.10^{-7}.t^2 + 1.1340782123112.10^{-3}.t \\ & + 2.070\end{aligned} \tag{3.1.1}$$

- Stress function [119]

$$\begin{aligned}\sigma(t) = {} & -1.2629384766298.10^{-39}.t^9 + 8.5820366306018.10^{-34}.t^8 - 1.913300488502.10^{-28}.t^7 \\ & + 2.0789371057655.10^{-23}.t^6 - 1.24789890007578.10^{-18}.t^5 + 4.2682280173128091.10^{-14}.t^4 \\ & - 8.07094539004.10^{-10}.t^3 + 7.631775481002.10^{-6}.t^2 - 2.81966349670.10^{-2}.t + 218.91\end{aligned} \tag{3.1.2}$$

1.1.2. Average functions

- Strain function [119]

$$\begin{aligned}\varepsilon(t) = {} & 2.913067377372647.10^{-40}.t^9 - 1.06287486879951.10^{-34}.t^8 + 1.62676364465.10^{-29}.t^7 \\ & - 1.35332812869675.10^{-24}.t^6 + 6.6258706347106649.10^{-20}.t^5 - 1.93082051629599.10^{-15}.t^4 \\ & + 3.225830232278.10^{-11}.t^3 - 2.80224085273862.10^{-7}.t^2 + 1.000835342972686.10^{-3}.t \\ & + 2.130\end{aligned} \tag{3.1.3}$$

- Stress function [119]

$$\sigma(t) = -1.354874822937.10^{-38}.t^9 + 4.9701415034543.10^{-33}.t^8 - 7.63138031480648.10^{-28}.t^7 + 6.3623058951283.10^{-23}.t^6 - 3.1226417988345039.10^{-18}.t^5 + 9.131325465772718.10^{-14}.t^4 - 1.5285804494941485.10^{-9}.t^3 + 1.3101398566887.10^{-5}.t^2 - 4.3003012174020541.10^{-2}.t + 205.68 \quad (3.1.4)$$

1.2. Elastic limit point

On table 3.1.2 is present a result after processing by stochastic method and average method. In fig. 3.1.2(a) shows the relationship between the change in relative strain versus the action time of the corrosion category, and in fig. 3.1.2(b) shows the relationship between the change in strength over time due to the corrosion effect of category.

Table 3.1.2 Result after processing

	Stochastic results									
time, [months]	0	9138	14769	24923	34892	46154	54000	64615	73108	77262
ε, [mm]	2.260	2.955	3.024	2.997	2.475	3.351	2.894	2.708	1.626	1.553
σ, [MPa]	292.80	293.90	276.50	283.50	281.70	282.40	286.80	283.90	233.90	232.90
	Average results									
time, [months]	0	9138	14769	24923	34892	46154	54000	64615	73108	77262
ε, [mm]	2.691	2.967	3.055	3.108	2.378	2.930	2.827	2.225	1.516	1.495
σ, [MPa]	284.00	277.30	276.60	264.10	265.00	265.10	288.50	248.00	211.20	232.20

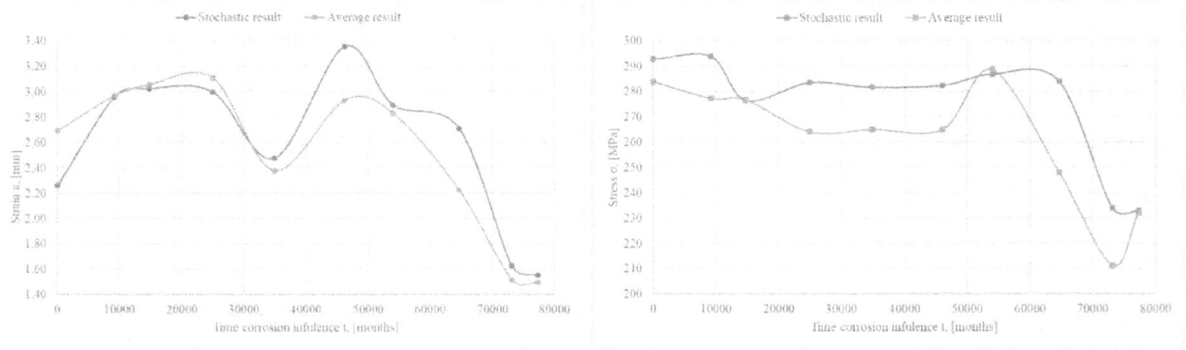

(a) (б)

Figure 3.1.2 (a) Graph of the variation of the strain deformation depending on the corrosion effect over time; (b) Graph of the change of stress strength depending on the corrosion effect over time;

Using the polynomial approximation [14-15,114-119] from fig. 3.1.2, functional dependence for the strain and stress point in dependence of the time of the influence of the corrosion (in months) is established in functions or equations.

1.2.1. Stochastic functions

- Strain function

$$\varepsilon(t) = 5.87434168267026.10^{-40}.t^9 - 2.08612709261823.10^{-34}.t^8 + 3.101605551012.10^{-29}.t^7 - 2.503133121768345.10^{-24}.t^6 + 1.1885876013233.10^{-19}\ t^5 - 3.36454227035366.10^{-15}.t^4 + 5.4847238628247926.10^{-11}.t^3 - 4.684011423204174.10^{-7}.t^2 + 1.657452896771.10^{-3}.t + 2.260 \quad (3.1.5)$$

- Stress function

$$\sigma(t) = 5.2605947800991.10^{-39}.t^9 - 1.8878048469246.10^{-34}.t^8 + 2.88648255011284.10^{-28}\ t^7 - 2.4491800005255.10^{-23}.t^6 + 1.255336106635019.10^{-18}.t^5 - 3.950699721462932.10^{-14}.t^4 + 7.372461550591.10^{-10}.t^3 - 7.322212773789.10^{-6}.t^2 + 2.8263424907545.10^{-2}.t + 292.84 \quad (3.1.6)$$

1.2.2. Average functions

- Strain function

$$\varepsilon(t) = 3.3872052818632.10^{-40}.t^9 - 1.23523951095922.10^{-34}.t^8 + 1.8892300816754.10^{-29}.t^7 - 1.5700583836327.10^{-24}.t^6 + 7.6748271971030336.10^{-20}.t^5 - 2.23063977110686.10^{-15}.t^4 + 3.7087584556614.10^{-11}.t^3 - 3.188004967825.10^{-7}.t^2 + 1.1031012729190133.10^{-3}.t + 2.691 \quad (3.1.7)$$

- Stress function

$$\sigma(t) = -\ 1.32088835777.10^{-38}.t^9 + 4.6733848517348498.10^{-33}.t^8 - 6.9081772383.10^{-28}.t^7 + 5.539806600969.10^{-23}.t^6 - 2.6184308755768133.10^{-18}.t^5 + 7.4103262193743949.10^{-14}.t^4 - 1.2137607947234564.10^{-9}.t^3 + 1.0385683547025486.10^{-5}.t^2 - 3.572353867740792.10^{-2}.t + 284.00 \quad (3.1.8)$$

1.3. Yield point

On table 3.1.3 is present a result after processing by stochastic method and average method. In fig. 3.1.3(a) shows the relationship between the change in relative strain versus the action time of the corrosion category, and in fig. 3.1.3(b) shows the relationship between the change in strength over time due to the corrosion effect of category.

Table 3.1.3 Result after processing [115, 117]

	Stochastic results									
time, [months]	0	9138	14769	24923	34892	46154	54000	64615	73108	77262
ε, [mm]	5.730	5.154	6.830	6.172	3.625	4.678	4.376	3.594	2.254	1.998
σ, [MPa]	358.67	332.90	329.90	339.80	354.10	337.40	352.80	331.90	327.40	312.90
	Average results									
time, [months]	0	9138	14769	24923	34892	46154	54000	64615	73108	77262
ε, [mm]	6.010	5.547	6.664	6.363	3.863	4.459	4.285	3.358	2.065	1.921
σ, [MPa]	344.79	326.20	347.10	352.10	329.80	337.20	361.10	305.20	280.00	317.10

Using the polynomial approximation [14-15,114-119] from fig. 3.1.3, functional dependence for the strain and stress point in dependence of the time of the influence of the corrosion (in months) is established in functions or equations.

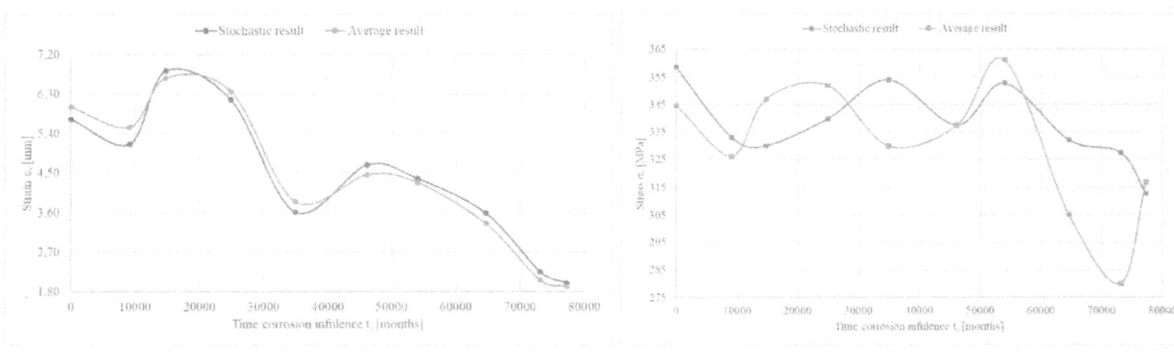

(a) (б)

Figure 3.1.3 (a) Graph of the variation of the strain deformation depending on the corrosion effect over time; [115]; (b) Graph of the change of stress strength depending on the corrosion effect over time [117];

1.3.1. Stochastic functions

- Strain function [115]

$$\varepsilon (t) = 7.310435890325.10^{-40}.t^{9} - 2.6011279426572.10^{-34}.t^{8} + 3.8527329197887.10^{-29}.t^{7} - 3.0661532491221.10^{-24}.t^{6} + 1.40941582847633.10^{-19}.t^{5} - 3.7314024722164327.10^{-15}.t^{4} \quad (3.1.9)$$

$$+ 5.32112101561890.10^{-11}.t^3 - 3.4750374028590.10^{-7}.t^2 + 7.0793882527168017.10^{-3}.t + 5.730$$

- Stress function [117]

$$\sigma (t) = - 1.5638535806745.10^{-38}.t^9 + 5.4803267401383.10^{-33}.t^8 - 8.020797363905.10^{-28}.t^7 + 6.355177840525386.10^{-23}.t^6 - 2.9547326021456.10^{-18}.t^5 + 8.17018162028724.10^{-14}.t^4 - 1.300333765472.10^{-9}.t^3 + 1.09247647690432.10^{-5}.t^2 - 3.941950471739783.10^{-2}.t + 358.67 \quad (3.1.10)$$

1.3.2. Average functions

- Strain function [115]

$$\varepsilon (t) = 6.3964855495508.10^{-40}.t^9 - 2.3121563161754.10^{-34}.t^8 + 3.4922209917485.10^{-29}.t^7 - 2.84880663712.10^{-24}.t^6 + 11.352862588713449.10^{-19}.t^5 - 3.749254196922094.10^{-15}.t^4 + 5.7394884544394706.10^{-11}.t^3 - 4.25622096088305.10^{-7}.t^2 + 1.12595672895339.10^{-3}.t + 6.010 \quad (3.1.11)$$

- Stress function [117]

$$\sigma (t) = - 7.864154863.10^{-39}.t^9 + 2.651313093144.10^{-33}.t^8 - 3.71480342046918.10^{-28}.t^7 + 2.821251600495785.10^{-23}.t^6 - 1.27570452962142.10^{-18}.t^5 + 3.56608575337818.10^{-14}.t^4 - 6.16162366206059.10^{-9}.t^3 + 6.12804951278145.10^{-5}.t^2 - 2.6491969753866387.10^{-2}.t + 344.79 \quad (3.1.12)$$

1.4. Lower yield point

On table 3.1.4 is present a result after processing by stochastic method and average method. In fig. 3.1.4(a) shows the relationship between the change in relative strain versus the action time of the corrosion category, and in fig. 3.1.4(b) shows the relationship between the change in strength over time due to the corrosion effect of category.

Using the polynomial approximation [14-15,114-119] from fig. 3.1.4, functional dependence for the strain and stress point in dependence of the time of the influence of the corrosion (in months) is established in functions or equations.

Table 3.1.4 Result after processing [118]

	Stochastic results									
time, [months]	0	9138	14769	24923	34892	46154	54000	64615	73108	77262
ε, [mm]	8.760	9.034	9.939	9.120	5.490	6.125	5.554	5.295	2.654	2.274
σ, [MPa]	363.43	347.40	366.40	405.30	357.90	372.90	369.80	345.50	377.90	363.90
	Average results									
time, [months]	0	9138	14769	24923	34892	46154	54000	64615	73108	77262
ε, [mm]	8.950	8.366	9.836	9.174	5.008	5.569	5.339	4.173	2.399	2.132
σ, [MPa]	359.19	343.70	365.90	376.90	352.90	358.60	379.90	316.10	295.70	339.40

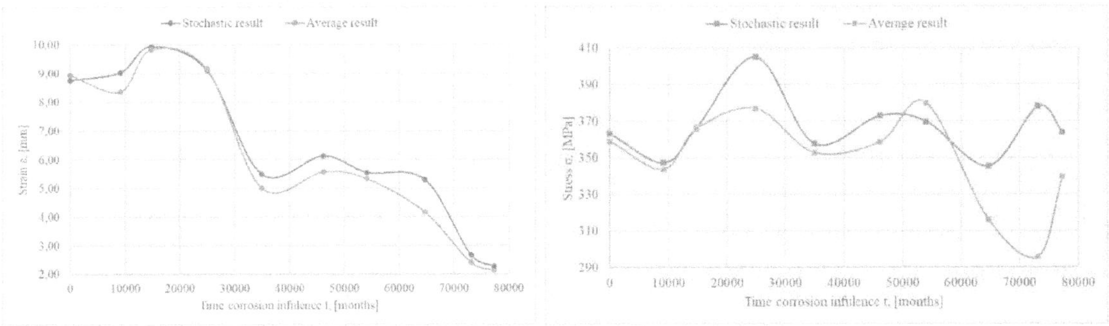

(a) (b)

Figure 3.1.4 (a) Graph of the variation of the strain deformation depending on the corrosion effect over time; [118]; (b) Graph of the change of stress strength depending on the corrosion effect over time [118];

1.4.1. Stochastic functions

- Strain function [118]

$$\varepsilon(t) = 1.222194748859.10^{-39}.t^{9} - 4.36180057173895.10^{-34}.t^{8} + 6.5152802381158.10^{-29}.t^{7} - 5.2743960539528.10^{-24}.t^{6} + 2.50164215671735.10^{-19}.t^{5} - 7.0066811630307258.10^{-15}.t^{4} + 1.109063431691.10^{-10}.t^{3} - 8.91364868125064.10^{-7}.t^{2} + 2.8167753759271242.10^{-3}.t + 8.760 \quad (3.1.13)$$

- Stress function [118]

$$\sigma(t) = 1.450092618775566.10^{-38}.t^{9} - 5.556362205558.10^{-33}.t^{8} + 8.903866838149.10^{-28}.t^{7} - 7.7287051791223.10^{-23}.t^{6} + 3.93014618143067.10^{-18}.t^{5} + 1.180225233780601.10^{-13}.t^{4} \quad (3.1.14)$$

$+ 1.997957823792752.10^{-9}.t^3 - 1.68699507413042.10^{-5}.t^2 + 5.265601380819774.10^{-2}.t + 363.43$

1.4.2. Average functions

- Strain function [118]

$$\varepsilon(t) = 9.7091351584243.10^{-40}.t^9 - 3.519601172432.10^{-34}.t^8 + 5.33481537669025.10^{-29}.t^7 - 4.371833192772737.10^{-24}.t^6 + 2.0888212525459.10^{-19}.t^5 - 5.838761504302115.10^{-15}.t^4 + 9.058322167491.10^{-11}.t^3 - 6.885552479570321.10^{-7}.t^2 + 1.9122463695163696.10^{-3}.t + 8.950 \quad (3.1.15)$$

- Stress function [118]

$$\sigma(t) = -7.48761686007275.10^{-39}.t^9 + 2.4767176729027.10^{-33}.t^8 - 3.3867252055.10^{-28}.t^7 + 2.49433814341168.10^{-23}.t^6 - 1.08738761786112.10^{-18}.t^5 + 2.92946844011926.10^{-14}.t^4 - 4.947901227320305.10^{-9}.t^3 - 4.95881581469996.10^{-6}.t^2 + 2.186848741130935.10^{-2}.t + 359.19 \quad (3.1.16)$$

1.5. Ultimate strength point

On table 3.1.5 is present a result after processing by stochastic method and average method. In fig. 3.1.5(a) shows the relationship between the change in relative strain versus the action time of the corrosion category, and in fig. 3.1.5(b) shows the relationship between the change in strength over time due to the corrosion effect of category.

Table 3.1.5 Result after processing [114]

	Stochastic results									
time, [months]	0	9138	14769	24923	34892	46154	54000	64615	73108	77262
ε, [mm]	28.540	28.370	23.790	19.730	15.270	13.120	11.250	8.090	6.070	3.430
σ, [MPa]	460.91	448.90	450.90	474.80	444.00	446.90	449.90	413.90	433.90	451.90
	Average results									
time, [months]	0	9138	14769	24923	34892	46154	54000	64615	73108	77262
ε, [mm]	29.600	23.410	23.440	19.440	12.470	12.250	12.00	8.890	4.550	3.550
σ, [MPa]	447.20	437.10	441.30	444.20	423.70	431.40	443.00	390.60	354.40	421.00

Using the polynomial approximation [14-15,114-119] from fig. 3.1.5, functional dependence for the strain and stress point in dependence of the time of the influence of the corrosion (in months) is established in functions or equations.

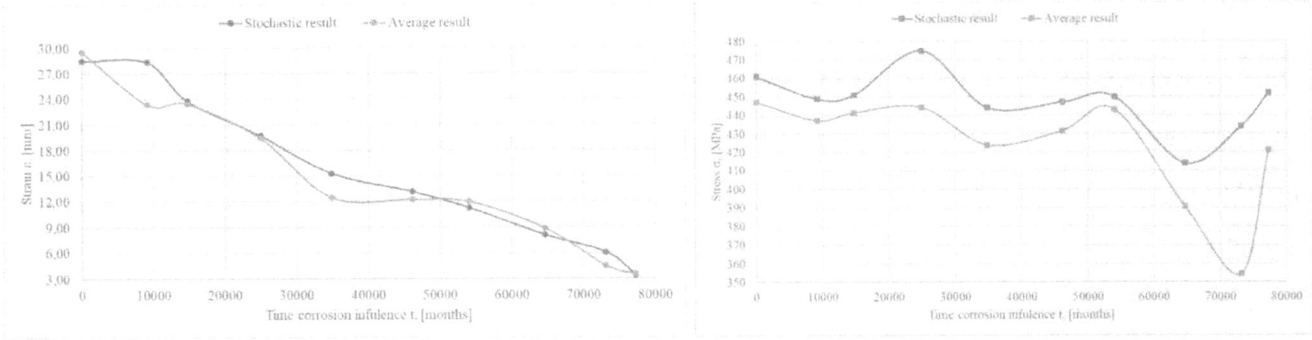

(a) (b)

Figure 3.1.5 (a) Graph of the variation of the strain deformation depending on the corrosion effect over time; [114]; (b) Graph of the change of stress strength depending on the corrosion effect over time [114];

1.5.1. Stochastic functions

- Strain function [114]

$$\varepsilon(t) = 6.4545951266348.10^{-40}.t^{9} - 2.4766871824102.10^{-34}.t^{8} + 4.0028292309866.10^{-29}.t^{7} - 3.5413290538715.10^{-24}.t^{6} + 1.865446792474038.10^{-19}.t^{5} - 5.95927744626138.10^{-15}.t^{4} + 1.1237426176627129.10^{-10}\, t^{3} - 1.14531822131232.10^{-6}\, t^{2} + 4.5138457495596.10^{-3}\, t + 28.540 \quad (3.1.17)$$

- Stress function [114]

$$\sigma(t) = 4.57795022128.10^{-39}.t^{9} - 1.9479633486225.10^{-33}.t^{8} + 3.42793718317145.10^{-28}.t^{7} - 3.237528228453.10^{-23}.t^{6} + 1.77786795216067.10^{-18}.t^{5} - 5.7327374297747193.10^{-14}.t^{4} + 1.038529733379187.10^{-9}.t^{3} - 9.38022344009419.10^{-6}.t^{2} + 3.090226699063908.10^{-2}.t + 460.91 \quad (3.1.18)$$

1.5.2. Average functions

- Strain function [114]

$$\varepsilon(t) = 9.2759471713087.10^{-40}.t^{9} - 3.371961984176.10^{-34}.t^{8} + 5.11789766476764.10^{-29}.t^{7} - 4.18536804797576.10^{-24}.t^{6} + 1.981766750034515.10^{-19}.t^{5} - 5.41548240923258.10^{-15}.t^{4} + 7.98210918936012.10^{-11}.t^{3} - 5.3239827695608047.10^{-7}.t^{2} + 5.1144095606624.10^{-4}.t + 29.600 \quad (3.1.19)$$

- Stress function [114]

$$\sigma(t) = 1.82789717263.10^{-40}.t^{9} - 1.5855582871437.10^{-34}.t^{8} + 4.15173713487605.10^{-29}.t^{7} - 5.0414938248317.10^{-24}.t^{6} + 3.236742807544566.10^{-19}.t^{5} - 1.12925080277007.10^{-14}.t^{4} + 2.0463322061035.10^{-10}.t^{3} - 1.6667522617780576.10^{-6}.t^{2} + 3.69575665085003.10^{-2}.t + 447.20 \quad (3.1.20)$$

1.6. Fracture point

On table 3.1.6 is present a result after processing by stochastic method and average method. In fig. 3.1.6(a) shows the relationship between the change in relative strain versus the action time of the corrosion category, and in fig. 3.1.6(b) shows the relationship between the change in strength over time due to the corrosion effect of category.

Table 3.1.6 Result after processing

	Stochastic results									
time, [months]	0	9138	14769	24923	34892	46154	54000	64615	73108	77262
ε, [mm]	40.410	36.080	28.360	23.420	17.600	16.640	12.990	10.230	6.750	5.170
σ, [MPa]	286.41	282.70	284.40	297.40	288.90	285.90	305.90	289.90	287.90	322.90
	Average results									
time, [months]	0	9138	14769	24923	34892	46154	54000	64615	73108	77262
ε, [mm]	38.800	29.180	28.250	23.020	15.120	14.370	14.160	10.280	5.630	4.310
σ, [MPa]	285.08	284.90	290.50	304.90	292.70	298.30	300.30	263.70	231.00	295.30

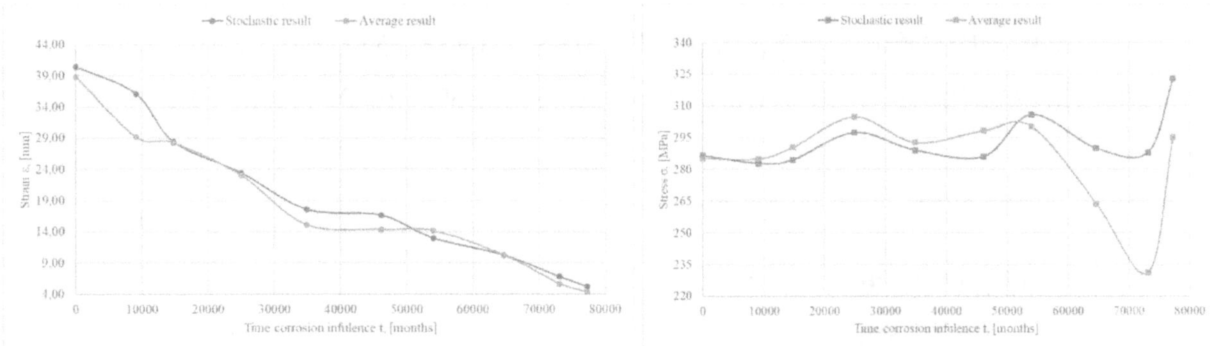

(a) (b)

Figure 3.1.6 (a) Graph of the variation of the strain deformation depending on the corrosion effect over time; (b) Graph of the change of stress strength depending on the corrosion effect over time;

Using the polynomial approximation [14-15,114-119] from fig. 3.1.6, functional dependence for the strain and stress point in dependence of the time of the influence of the corrosion (in months) is established in functions or equations.

1.6.1. Stochastic functions

- Strain function

$$\varepsilon(t) = -7.48761686007275.10^{-39}.t^9 + 2.4767176729.10^{-33}.t^8 - 3.38672520550253.10^{-28}.t^7 + 2.494338143411675.10^{-23}.t^6 - 1.08738761786112.10^{-18}.t^5 + 2.92946844011926.10^{-14}.t^4 - 4.9479012273203055.10^{-10}.t^3 + 4.95881581469996.10^{-6}.t^2 - 2.186848741130935.10^{-2}.t + 40.410 \quad (3.1.21)$$

- Stress function

$$\sigma(t) = -5.1399397335.10^{-39}.t^9 + 1.70568917810045.10^{-33}.t^8 - 2.3147670328044.10^{-28}.t^7 + 1.6545228273015.10^{-23}.t^6 - 6.69353168567666.10^{-19}.t^5 + 1.5386185089679071.10^{-14}.t^4 - 1.9398680389107966.10^{-10}.t^3 + 1.2773466121268.10^{-6}.t^2 - 3.881372047517508.10^{-3}.t + 286.41 \quad (3.1.22)$$

1.6.2. Average functions

- Strain function

$$\varepsilon(t) = 7.7770339231175.10^{-40}.t^9 - 2.8620284054562.10^{-34}.t^8 + 4.3936708981436.10^{-29}.t^7 - 3.62668245607664.10^{-24}.t^6 + 1.7260583943026267.10^{-19}.t^5 - 4.7008011303508.10^{-15}.t^4 + 6.7650874288029428.10^{-11}.t^3 - 4.072791743155147.10^{-7}.t^2 - 3.89686917878038.10^{-4}.t + 38.800 \quad (3.1.23)$$

- Stress function

$$\sigma(t) = 4.70409729295305.10^{-39}.t^9 - 1.717570273329.10^{-33}.t^8 + 2.6639768627167.10^{-28}.t^7 - 2.26973335152976.10^{-23}.t^6 + 1.14631961264479.10^{-18}.t^5 - 3.452275493301128.10^{-14}.t^4 + 5.91207043584759.10^{-10}.t^3 - 5.10391051110766.10^{-6}.t^2 + 1.6901868314524074.10^{-3}.t + 285.08 \quad (3.1.24)$$

1.7. Young's modulus (E-modulus)

On table 3.1.7 is present a result is presented after processing, as the given values of the E-modulus are calculated as the tangent of the angle that the deformation curve makes with the abscissa axis of the σ-ε diagram and subsequently (the obtained values) are processed using the stochastic method and average method. Fig. 3.3.7 shows the relationship between the change in E-modulus and the time of action of the corrosion category.

Table 3.1.7 Result after processing

	Stochastic results									
time, [months]	0	9138	14769	24923	34892	46154	54000	64615	73108	77262
E, [GPa]	211.50	168.10	153.50	148.70	156.20	151.90	131.60	153.50	184.50	232.00
	Average results									
time, [months]	0	9138	14769	24923	34892	46154	54000	64615	73108	77262
E, [GPa]	193.10	166.20	159.10	143.60	156.30	164.40	136.50	157.50	181.20	206.60

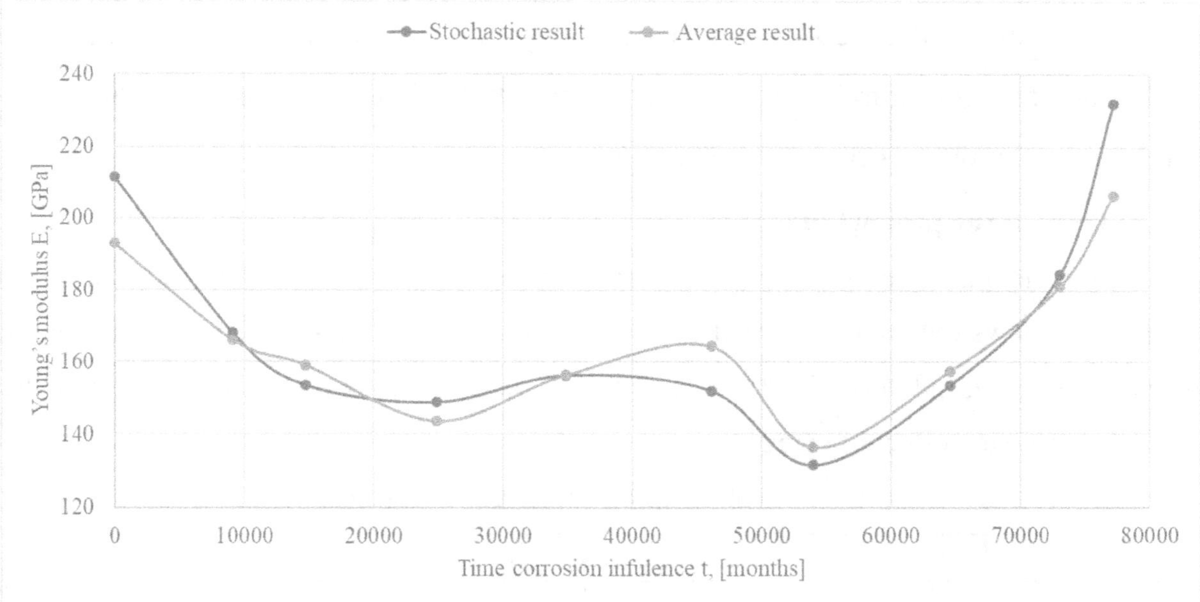

Figure 3.1.7 Graph of the variation of the Young's modulus (E-modulus) depending on the corrosion effect over time;

Using the polynomial approximation [14-15,114-119] from fig. 3.3.7, functional dependence for the E-modulus in dependence of the time of the influence of the corrosion (in months) is established in functions or equations.

1.7.1. Stochastic functions

$$E(t) = 7.40959020263794.10^{-39}.t^9 - 2.51056192868756.10^{-33}.t^8 + 3.54362261975.10^{-28}.t^7 - 2.7048099822994.10^{-23}.t^6 + 1.214067989473427.10^{-18}.t^5 - 3.263220762815896.10^{-14}.t^4 + 5.0767575912875.10^{-10}.t^3 - 3.983056131294.10^{-6}.t^2 + 7.2200315076118779.10^{-3}.t + 211.50 \quad (3.1.25)$$

1.7.2. Average functions

$$E(t) = 9.6252852979736.10^{-39}.t^9 - 3.2225824772648.10^{-33}.t^8 + 4.453178729197.10^{-28}\ t^7 - 3.281472358756075.10^{-23}.t^6 + 1.3923238485823.10^{-18}.t^5 - 3.4337073074345.10^{-14}.t^4 + 4.73231969985348.10^{-10}.t^3 - 3.2188053756442576.10^{-6}.t^2 + 5.2942365104152.10^{-3}.t + 193.10 \quad (3.1.26)$$

2. Corrosion categoryC2

According to ISO 12944-5, corrosion category C2 is for environments with low pollution levels, such as agricultural areas. Unheated buildings where condensation can occur, such as warehouses, depots, sports facilities and others.

2.1. Point of the proportional limit

On table 3.2.1 is present a result after processing by stochastic method and average method. In fig. 3.2.1(a) shows the relationship between the change in relative strain versus the action time of the corrosion category, and in fig. 3.2.1(b) shows the relationship between the change in strength over time due to the corrosion effect of category.

Table 3.2.1 Result after processing [119]

	Stochastic results									
time, [months]	0	475	768	1296	1814	2400	2808	3360	3802	4018
ε, [mm]	2.070	2.521	2.727	2.595	2.151	2.87	2.432	2.296	1.414	1.331
σ, [MPa]	218.91	211.90	209.30	193.00	224.00	218.00	213.40	234.90	173.90	205.90
	Average results									
time, [months]	0	475	768	1296	1814	2400	2808	3360	3802	4018
ε, [mm]	2.130	2.543	2.619	2.664	2.039	2.511	2.423	1.907	1.299	1.282
σ, [MPa]	205.68	211.30	208.40	191.30	212.40	206.40	220.50	200.20	156.90	176.60

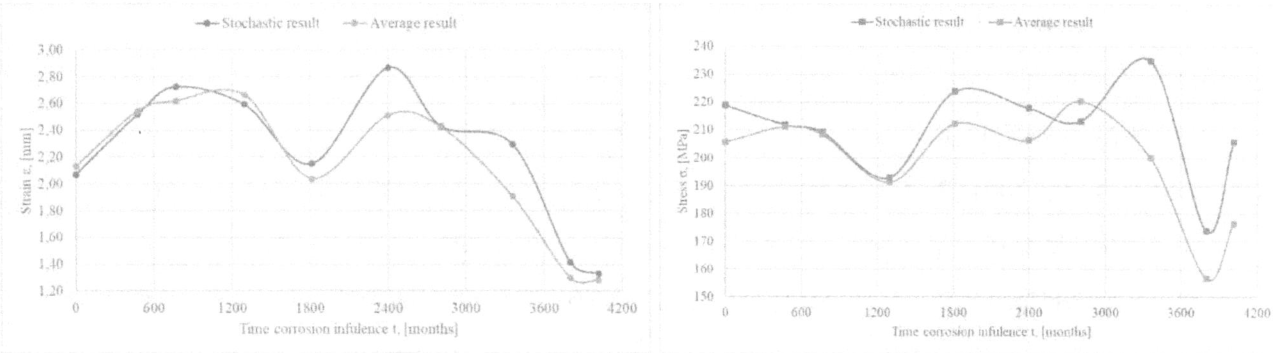

(a) (b)

Figure 3.2.1 (a) Graph of the variation of the strain deformation depending on the corrosion effect over time; [119]; (b) Graph of the change of stress strength depending on the corrosion effect over time [119];

Using the polynomial approximation [14-15,114-119] from fig. 3.2.1, functional dependence for the strain and stress point in dependence of the time of the influence of the corrosion (in months) is established in functions or equations.

2.1.1. Stochastic functions

- Strain function [119]

$$\varepsilon(t) = 1.70891186723.10^{-28}.t^9 - 3.134594280939.10^{-24}.t^8 + 2.402602509830143.10^{-20}.t^7 -9.9701197493765.10^{-17}.t^6 + 2.425317943544567.10^{-13}.t^5 - 3.498467184965793.10^{-10}.t^4 + 2.88418182301801.10^{-7}.t^3 - 1.2338745325756.10^{-4}.t^2 + 2.178294462287816.10^{-2}.t + 2.070 \quad (3.2.1)$$

- Stress function [119]

$$\sigma(t) = -4.619585151279.10^{-28}.t^9 + 1.6187538365857.10^{-23}.t^8 - 1.8709452768788.10^{-19}.t^7 + 1.05548896344624.10^{-15}.t^6 - 3.29151747379097.10^{-12}.t^5 + 5.8506781305396841.10^{-9}.t^4 - 5.75046565557846.10^{-6}.t^3 + 2.82662540824356.10^{-3}.t^2 - 5.428846463623.10^{-1}.t + 218.91 \quad (3.2.2)$$

2.1.2. Average functions

- Strain function [119]

$$\varepsilon(t) = 1.04727134731596.10^{-28}.t^9 - 1.987086765145.10^{-24}.t^8 + 1.581545211298.10^{-20}.t^7 - 6.841927666429.10^{-17}.t^6 + 1.741932385593324.10^{-13}.t^5 - 2.6395898589470038.10^{-10}.t^4 \quad (3.2.3)$$

$+ 2.29315381185895.10^{-7}.t^3 - 1.03581614286303.10^{-4}.t^2 + 1.92361777980328.10^{-2}.t + 2.130$

- Stress function [119]

$$\sigma(t) = -4.87333964390123.10^{-27}.t^9 + 9.296029575779.10^{-23}.t^8 - 7.4221856032545.10^{-19}.t^7 + 3.2176681781621.10^{-15}.t^6 - 8.211902904245.10^{-12}.t^5 + 1.2486637237217132.10^{-8}.t^4 - 1.08688064376598.10^{-5}.t^3 + 4.84371604979252.10^{-3}.t^2 - 8.26609628597139.10^{-1}.t + 205.68 \quad (3.2.4)$$

2.2. Elastic limit point

On table 3.2.2 is present a result after processing by stochastic method and average method. In fig. 3.2.2(a) shows the relationship between the change in relative strain versus the action time of the corrosion category, and in fig. 3.2.2(b) shows the relationship between the change in strength over time due to the corrosion effect of category.

Table 3.2.2 Result after processing

	Stochastic results									
time, [months]	0	475	768	1296	1814	2400	2808	3360	3802	4018
ε, [mm]	2.260	2.955	3.024	2.997	2.475	3.351	2.894	2.708	1.626	1.553
σ, [MPa]	292.80	293.90	276.50	283.50	281.70	282.40	286.80	283.90	233.90	232.90
	Average results									
time, [months]	0	475	768	1296	1814	2400	2808	3360	3802	4018
ε, [mm]	2.691	2.967	3.055	3.108	2.378	2.930	2.827	2.225	1.516	1.495
σ, [MPa]	284.00	277.30	276.60	264.10	265.00	265.10	288.50	248.00	211.20	232.20

Using the polynomial approximation [14-15,114-119] from fig. 3.2.2, functional dependence for the strain and stress point in dependence of the time of the influence of the corrosion (in months) is established in functions or equations.

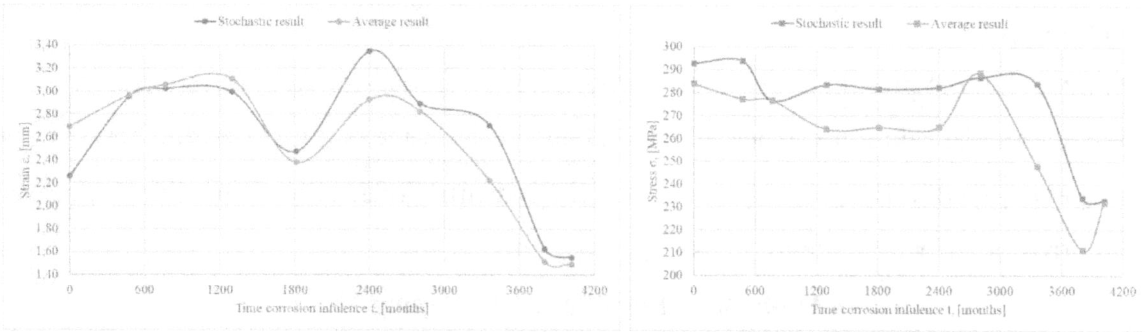

(a) (b)

Figure 3.2.2 (a) Graph of the variation of the strain deformation depending on the corrosion effect over time; (b) Graph of the change of stress strength depending on the corrosion effect over time;

2.2.1. Stochastic functions

- Strain function

$$\varepsilon(t) = 2.1107061494344.10^{-28}.t^9 - 3.897903899205.10^{-24}.t^8 + 3.0136658426821.10^{-20}.t^7 - 1.26475471570757.10^{-16}.t^6 + 3.12292913320863.10^{-13}.t^5 - 4.5968364059698.10^{-10}.t^4 + 3.896567193118965.10^{-7}.t^3 - 1.730342413615535.10^{-4}.t^2 + 3.18383246236953.10^{-2}.t + 2.260 \quad (3.2.5)$$

- Stress function

$$\sigma(t) = 1.8888646446163.10^{-27}.t^9 - 3.525136096215.10^{-23}.t^8 + 2.803116577061.10^{-19}.t^7 - 1.2369302099326125.10^{-15}.t^6 + 3.2971058146922466.10^{-12}.t^5 - 5.39619672355977.10^{-9}.t^4 + 5.2366690518791.10^{-6}.t^3 - 2.70454816381798.10^{-3}.t^2 + 0.54280612919114.t + 292.84 \quad (3.2.6)$$

2.2.2. Average functions

- Strain function

$$\varepsilon(t) = 1.21772180481597.10^{-28}.t^9 - 2.309317194183.10^{-24}.t^8 + 1.8367054759104.10^{-20}.t^7 - 7.9375849783326.10^{-17}.t^6 + 2.01768655215336.10^{-13}.t^5 - 3.0494390689674294.10^{-10}.t^4 + 2.63642080222884.10^{-7}.t^3 - 1.17838564549782.10^{-4}.t^2 + 2.12008484772.10^{-2}.t + 2.691 \quad (3.2.7)$$

- Stress function

$$\sigma(t) = - 4.747796557436.10^{-37}.t^9 + 8.7350079376275.10^{-23}.t^8 - 6.71425796856496.10^{-19}.t^7 + 2.79981064094487.10^{-15}.t^6 - 6.8813087332253765.10^{-12}.t^5 + 1.01264884883945.10^{-8}.t^4 - 8.6246270865341354.10^{-6}.t^3 + 3.83723018824857.10^{-3}.t^2 - 0.68628198775214755.t + 284.00 \quad (3.2.8)$$

2.3. Yield point

On table 3.2.3 is present a result after processing by stochastic method and average method. In fig. 3.2.3(a) shows the relationship between the change in relative strain versus the action time of the corrosion category, and in fig. 3.2.3(b) shows the relationship between the change in strength over time due to the corrosion effect of category.

Table 3.2.3 Result after processing [115, 117]

	Stochastic results									
time, [months]	0	475	768	1296	1814	2400	2808	3360	3802	4018
ε, [mm]	5.730	5.154	6.830	6.172	3.625	4.678	4.376	3.594	2.254	1.998
σ, [MPa]	358.67	332.90	329.90	339.80	354.10	337.40	352.80	331.90	327.40	312.90
	Average results									
time, [months]	0	475	768	1296	1814	2400	2808	3360	3802	4018
ε, [mm]	6.010	5.547	6.664	6.363	3.863	4.459	4.285	3.358	2.065	1.921
σ, [MPa]	344.79	326.20	347.10	352.10	329.80	337.20	361.10	305.20	280.00	317.10

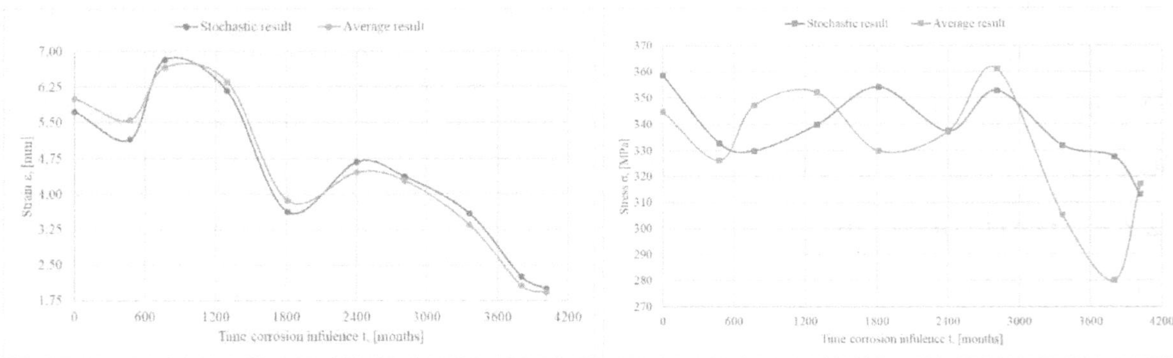

(a) (b)

Figure 3.2.3 (a) Graph of the variation of the strain deformation depending on the corrosion effect over time; [115]; (b) Graph of the change of stress strength depending on the corrosion effect over time [117];

Using the polynomial approximation [14-15,114-119] from fig. 3.2.3, functional dependence for the strain and stress point in dependence of the time of the influence of the corrosion (in months) is established in functions or equations.

2.3.1. Stochastic functions

- Strain function [115]

$$\varepsilon(t) = 2.6291133102476.10^{-28}.t^9 - 4.864755115792.10^{-24}.t^8 + 3.747147161906.10^{-20}.t^7 -1.5508094948814.10^{-16}.t^6 +3.707135867551699.10^{-13}.t^5 - 5.1040218297501278.10^{-10}.t^4 + 3.78531690318262.10^{-7}.t^3 - 1.2857923202246.10^{-4}.t^2 + 1.3630684360611892.10^{-2}.t + 5.730 \quad (3.2.9)$$

- Stress function [117]

$$\sigma(t) = - 5.6196876179019.10^{-27}.t^9 + 1.0240882751.10^{-22}.t^8 - 7.79398384854764.10^{-19}.t^7 + 3.2112654063577.10^{-15}.t^6 - 7.7636858494722047.10^{-12}.t^5 + 1.11628902375046.10^{-8}.t^4 - 9.238228385674415.10^{-6}.t^3 + 4.035833852196446.10^{-3}.t^2 - 7.57265539574768.10^{-1}.t + 358.67 \quad (3.2.10)$$

2.3.2. Average functions

- Strain function [115]

$$\varepsilon(t) = 2.3010571320007.10^{-28}.t^9 - 4.325476955956.10^{-24}.t^8 + 3.39740715988153.10^{-20}.t^7 - 1.44124083686674.10^{-16}.t^6 + 3.55921796273326.10^{-13}.t^5 - 5.12949751555918.10^{-10}.t^4 + 4.083558107115.10^{-7}.t^3 - 1.5748642820118729.10^{-4}.t^2 + 2.1668826777941359.10^{-2}.t + 6.010 \quad (3.2.11)$$

- Stress function [117]

$$\sigma(t) = - 2.8230227697.10^{-27}.t^9 + 4.948523263888.10^{-22}.t^8 - 3.6048437037288.10^{-19}.t^7 +1.4233566079976.10^{-15}.t^6 - 3.34610185472458.10^{-12}.t^5 + 4.8631947476896104.10^{-9}.t^4 - 4.369525303617237.10^{-6}.t^3 + 2.260301892656886.10^{-3}.t^2 - 5.083047923479.10^{-1}.t + 344.79 \quad (3.2.12)$$

2.4. Lower yield point

On table 3.2.4 is present a result after processing by stochastic method and average method. In fig. 3.2.4(a) shows the relationship between the change in relative strain versus the action time of the corrosion category, and in fig. 3.2.4(b) shows the relationship between the change in strength over time due to the corrosion effect of category.

Using the polynomial approximation [14-15,114-119] from fig. 3.2.4, functional dependence for the strain and stress point in dependence of the time of the influence of the corrosion (in months) is established in functions or equations.

Table 3.2.4 Result after processing [118]

	Stochastic results									
time, [months]	0	475	768	1296	1814	2400	2808	3360	3802	4018
ε, [mm]	8.760	9.034	9.939	9.12	5.49	6.125	5.554	5.295	2.654	2.274
σ, [MPa]	363.43	347.40	366.40	405.30	357.90	372.90	369.80	345.50	377.90	363.90
	Average results									
time, [months]	0	475	768	1296	1814	2400	2808	3360	3802	4018
ε, [mm]	8.950	8.366	9.836	9.174	5.008	5.569	5.339	4.173	2.399	2.132
σ, [MPa]	359.19	343.70	365.90	376.90	352.90	358.60	379.90	316.10	295.70	339.40

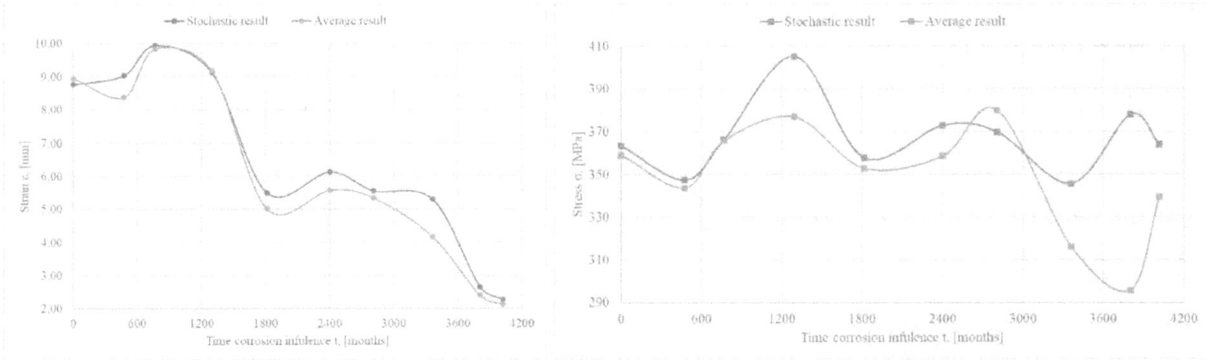

(a) (b)

Figure 3.2.4 (a) Graph of the variation of the strain deformation depending on the corrosion effect over time; [118]; (b) Graph of the change of stress strength depending on the corrosion effect over time [118];

2.4.1. Stochastic functions

- Strain function [118]

$$\varepsilon(t) = -2.8230227697067.10^{-27}.t^{9} + 4.948523263888.10^{-23}.t^{8} - 3.6048437037288.10^{-19}.t^{7} + 1.4233566079976.10^{-15}.t^{6} - 3.34610185472458.10^{-12}.t^{5} + 4.86319474768961.10^{-9}.t^{4} - 4.369525303617237.10^{-6}.t^{3} + 2.260301892656886.10^{-3}.t^{2} - 5.0830479234795289.10^{-1}.t + 8.760 \quad (3.2.13)$$

- Stress function [118]

$$\sigma(t) = 5.2201584797068.10^{-27}.t^9 - 1.040049268666.10^{-22}.t^8 + 8.66602504998687.10^{-19}.t^7 - 3.91136027906726.10^{-15}.t^6 + 1.03421478356258.10^{-11}.t^5 - 1.61490949391139.10^{-8}.t^4 + 1.4215103278623326.10^{-5}.t^3 - 6.2410116668287219.10^{-3}.t^2 + 1.0128744212189265.t + 363.43 \quad (3.2.14)$$

2.4.1. Average functions

- Strain function [118]

$$\varepsilon(t) = 3.493340809956.10^{-28}.t^9 - 6.585416538379.10^{-24}.t^8 + 5.19082807042238.10^{-20}.t^7 - 2.2121100546254.10^{-16}.t^6 + 5.49629049846198.10^{-13}.t^5 - 7.9894290350258694.10^{-10}.t^4 + 6.4457733562884127.10^{-7}.t^3 - 2.54807650719059.10^{-4}.t^2 + 3.6803663573919995.10^{-2}.t + 8.950 \quad (3.2.15)$$

- Stress function [118]

$$\sigma(t) = -2.6868797049072.10^{-27}.t^9 + 4.620688312862.10^{-23}.t^8 - 3.284790643362.10^{-19}.t^7 + 1.25762287388998.10^{-15}.t^6 - 2.849885143548005.10^{-12}.t^5 + 3.99120544262438.10^{-9}.t^4 - 3.5052826574178763.10^{-6}.t^3 + 1.8274650973704361.10^{-3}.t^2 - 4.1932628867208827.t + 359.19 \quad (3.2.16)$$

2.5. Ultimate strength point

On table 3.2.5 is present a result after processing by stochastic method and average method. In fig. 3.2.5(a) shows the relationship between the change in relative strain versus the action time of the corrosion category, and in fig. 3.2.5(b) shows the relationship between the change in strength over time due to the corrosion effect of category.

Table 3.2.5 Result after processing [114]

	Stochastic results									
time, [months]	0	475	768	1296	1814	2400	2808	3360	3802	4018
ε, [mm]	28.540	28.370	23.790	19.730	15.270	13.120	11.250	8.090	6.070	3.430
σ, [MPa]	460.91	448.90	450.90	474.80	444.00	446.90	449.90	413.90	433.90	451.90
	Average results									
time, [months]	0	475	768	1296	1814	2400	2808	3360	3802	4018
ε, [mm]	29.600	23.410	23.440	19.440	12.470	12.250	12.000	8.890	4.550	3.550
σ, [MPa]	447.20	437.10	441.30	444.20	423.70	431.40	443.00	390.60	354.40	421.00

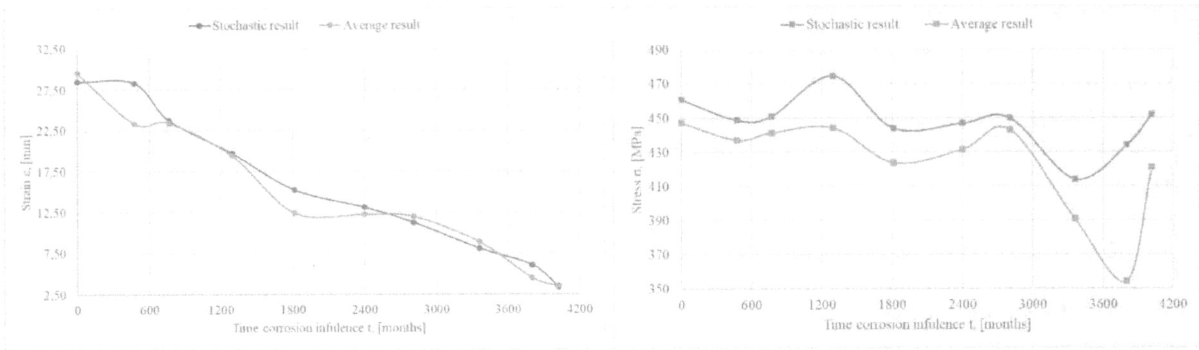

(a) (b)

Figure 3.2.5 (a) Graph of the variation of the strain deformation depending on the corrosion effect over time; [114]; (b) Graph of the change of stress strength depending on the corrosion effect over time [114];

Using the polynomial approximation [14-15,114-119] from fig. 3.2.5, functional dependence for the strain and stress point in dependence of the time of the influence of the corrosion (in months) is established in functions or equations.

2.5.1. Stochastic functions

- Strain function [114]

$$\varepsilon(t) = 2.3240385748615.10^{-28}.t^{9} - 4.6367828710858.10^{-24}.t^{8} + 3.8965212710032.10^{-20}.t^{7} - 1.792384915142448.10^{-16}.t^{6} + 4.908969628913.10^{-13}.t^{5} - 8.1531857850936906.10^{-10}.t^{4} + 7.9929593880578318.10^{-7}.t^{3} - 4.23502955999745.10^{-6}.t^{2} + 8.6765400965155.10^{-3}.t + 28.540 \quad (3.2.17)$$

- Stress function [114]

$$\sigma(t) = 1.6514991598233.10^{-27}.t^{9} - 3.65226407822.10^{-23}.t^{8} + 3.3407191392583.10^{-19}.t^{7} - 1.6401492675665.10^{-15}.t^{6} + 4.6822930399357427.10^{-12}.t^{5} - 7.8492691677063125.10^{-9}.t^{4} + 7.392763902204428.10^{-6}.t^{3} - 3.471591769618.10^{-3}.t^{2} + 5.9460819178219415.10^{-1}.t + 460.91 \quad (3.2.18)$$

2.5.2. Average functions

- Strain function [114]

$$\varepsilon(t) = 3.339312354582.10^{-28}.t^{9} - 6.31269212997427.10^{-24}.t^{8} + 4.9825740974486.10^{-20}.t^{7} - 2.118967474802146.10^{-16}.t^{6} + 5.217620246493.10^{-13}.t^{5} - 7.4146846024057.10^{-10}.t^{4} \quad (3.2.19)$$

$+ 5.683660245369694.10^{-7}.t^3 - 1.9718042309595.10^{-6}.t^2 + 9.880023331196689.10^{-3}.t$
$+ 29.600$

- Stress function [114]

$$\sigma(t) = 1.65149915982334.10^{-27}.t^9 - 3.65226407822.10^{-23}.t^8 + 3.3407191392583.10^{-19}.t^7 - 1.6401492675665.10^{-15}.t^6 + 4.6822930399357427.10^{-12}.t^5 - 7.8492691677063125.10^{-9}.t^4 + 7.392763902204428.10^{-6}.t^3 - 3.471591769618.10^{-3}.t^2 + 5.9460819178219415.10^{-1}.t + 447.20 \quad (3.2.20)$$

2.6. Fracture point

On table 3.2.6 is present a result after processing by stochastic method and average method. In fig. 3.2.6(a) shows the relationship between the change in relative strain versus the action time of the corrosion category, and in fig. 3.2.6(b) shows the relationship between the change in strength over time due to the corrosion effect of category.

Table 3.2.6 Result after processing

	Stochastic results									
time, [months]	0	475	768	1296	1814	2400	2808	3360	3802	4018
ε, [mm]	40.410	36.080	28.360	23.420	17.600	16.640	12.990	10.230	6.750	5.170
σ, [MPa]	286.41	282.70	284.40	297.40	288.90	285.90	305.90	289.90	287.90	322.90
	Average results									
time, [months]	0	475	768	1296	1814	2400	2808	3360	3802	4018
ε, [mm]	38.800	29.180	28.250	23.020	15.120	14.370	14.160	10.280	5.630	4.310
σ, [MPa]	285.08	284.90	290.50	304.90	292.70	298.30	300.30	263.70	231.00	295.30

Using the polynomial approximation [14-15,114-119] from fig. 3.2.6, functional dependence for the strain and stress point in dependence of the time of the influence of the corrosion (in months) is established in functions or equations.

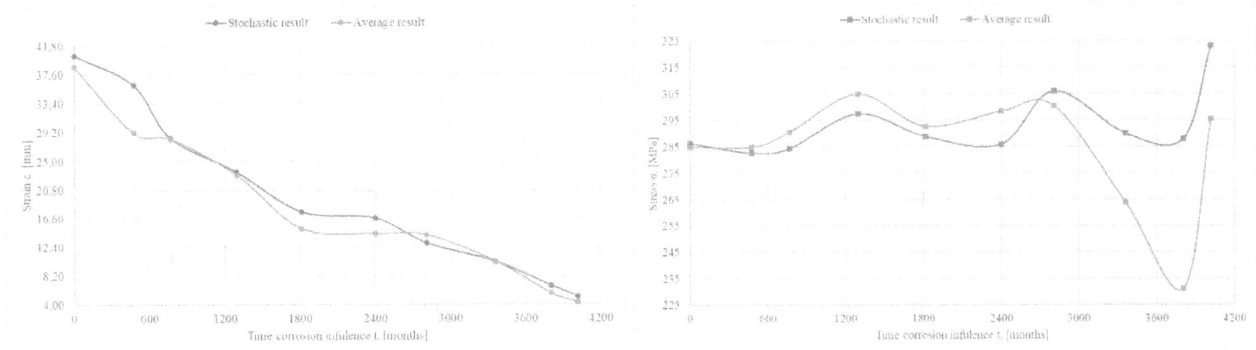

(a) (b)

Figure 3.2.6 (a) Graph of the variation of the strain deformation depending on the corrosion effect over time; (b) Graph of the change of stress strength depending on the corrosion effect over time;

2.6.1. Stochastic functions

- Strain function

$$\varepsilon(t) = 8.4235255383815.10^{-28}.t^9 - 1.59556545716676.10^{-23}.t^8 + 1.271705364029.10^{-19}.t^7 - 5.5408201380323.10^{-16}.t^6 + 1.4346466057598757.10^{-12}.t^5 - 2.2454337740304.10^{-9}.t^4 + 2.06033426548842.10^{-6}.t^3 - 1.00402572624913.10^{-3}.t^2 + 1.8257004549032085.10^{-1}.t + 40.410 \quad (3.2.21)$$

- Stress function

$$\sigma(t) = -1.84492775480218.10^{-27}.t^9 + 3.183212734215.10^{-23}.t^8 - 2.245877456509.10^{-19}.t^7 + 8.344814435947 98.10^{-16}.t^6 - 1.75464160159846.10^{-12}.t^5 + 2.095703390361489.10^{-9}.t^4 - 1.37230919293160.10^{-6}.t^3 + 4.6918080502449696.10^{-4}.t^2 - 7.40892879612436.10^{-2}.t + 286.41 \quad (3.2.22)$$

2.6.2. Average functions

- Strain function

$$\varepsilon(t) = 2.801494797964.10^{-28}.t^9 - 5.361309596593.10^{-24}.t^8 + 4.2799887154975.10^{-20}.t^7 - 1.837140075902037.10^{-16}.t^6 + 4.54684463159348.10^{-13}.t^5 - 6.43965431965088.10^{-10}.t^4 + 4.81992289830675.10^{-7}.t^3 - 1.50970256739026.10^{-4}.t^2 + 7.4435020215388933.10^{-3}.t + 38.800 \quad (3.2.23)$$

- Stress function

$$\sigma(t) = 1.691060279226.10^{-27}.t^9 - 3.21116102473647.10^{-23}.t^8 + 2.5902111601277.10^{-19}.t^7 - 1.1476965957758.10^{-15}.t^6 + 3.01435540259855.10^{-12}.t^5 - 4.7208311489183717.10^{-9}.t^4 \quad (3.2.24)$$

$+ 4.2040045911634479.10^{-6}.t^3 - 1.8872099559.10^{-4}.t^2 + 3.2495455376730187.10^{-2}.t + 285.08$

2.7. Young's modulus (E-modulus)

On table 3.2.7 is present a result is presented after processing, as the given values of the E-modulus are calculated as the tangent of the angle that the deformation curve makes with the abscissa axis of the σ-ε diagram and subsequently (the obtained values) are processed using the stochastic method and average method.

Table 3.2.7 Result after processing

	Stochastic results									
time, [months]	0	475	768	1296	1814	2400	2808	3360	3802	4018
E, [GPa]	211.50	168.10	153.50	148.70	156.20	151.90	131.60	153.50	184.50	232.00
	Average results									
time, [months]	0	475	768	1296	1814	2400	2808	3360	3802	4018
E, [GPa]	193.10	166.20	159.10	143.60	156.30	164.40	136.50	157.50	181.20	206.60

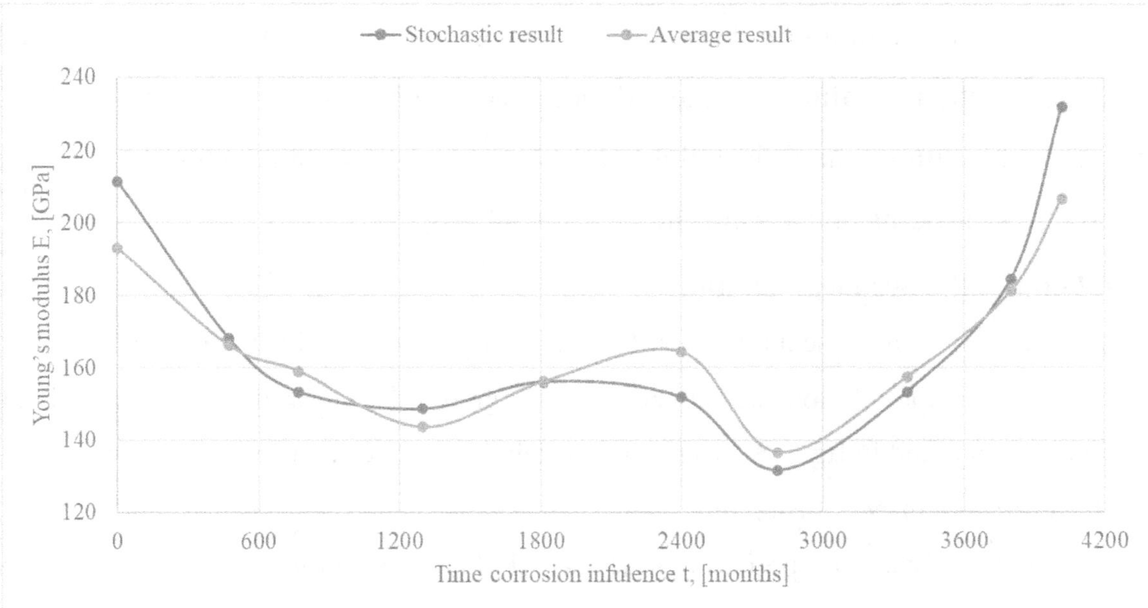

Figure 3.2.7 Graph of the variation of the Young's modulus (E-modulus) depending on the corrosion effect over time;

Fig. 3.3.7 shows the relationship between the change in E-modulus and the time of action of the corrosion category.

Using the polynomial approximation [14-15,114-119] from fig. 3.3.7, functional dependence for the E-modulus in dependence of the time of the influence of the corrosion (in months) is established in functions or equations.

2.7.1. Stochastic functions

$$E(t) = 2.661287038142.10^{-27}.t^9 - 4.68863814109.10^{-23}.t^8 + 3.4410387210065.10^{-19}.t^7 - 1.3656228418058.10^{-15}.t^6 + 3.1869274412509.10^{-12}\,t^5 - 4.4533991943847887.10^{-9}.t^4 + 3.60180534542143.10^{-6}.t^3 - 1.4687576272696.10^{-3}.t^2 + 0.13799565001668757.t + 211.50 \quad (3.2.25)$$

2.7.2. Average functions

$$E(t) = 3.4556311752095.10^{-27}.t^9 - 6.01559500832.10^{-23}.t^8 + 4.321999538062.10^{-19}.t^7 - 1.6557478901922.10^{-15}.t^6 + 3.6520330832026952.10^{-12}.t^5 - 4.6812913413190067.10^{-9}.t^4 + 43.352684240937.10^{-6}.t^3 - 1.1845209537897959.10^{-3}.t^2 + 0.10074858439230133.t + 193.10 \quad (3.2.26)$$

3. Corrosion category C3

It exists in urban and industrial regions with moderate sulfur dioxide pollution and in coastal areas with low salinity. In the indoor area of a building, it is available in the production part with high humidity and air pollution, such as industrial plants, laundry rooms, breweries, food production rooms, etc., according to ISO 12944-5.

3.1. Point of the proportional limit

On table 3.3.1 is present a result after processing by stochastic method and average method. In fig. 3.3.1(a) shows the relationship between the change in relative strain versus the action time of the corrosion category, and in fig. 3.3.1(b) shows the relationship between the change in strength over time due to the corrosion effect of category.

Using the polynomial approximation [14-15,114-119] from fig. 3.3.1, functional dependence for the strain and stress point in dependence of the time of the influence of the corrosion (in months) is established in functions or equations.

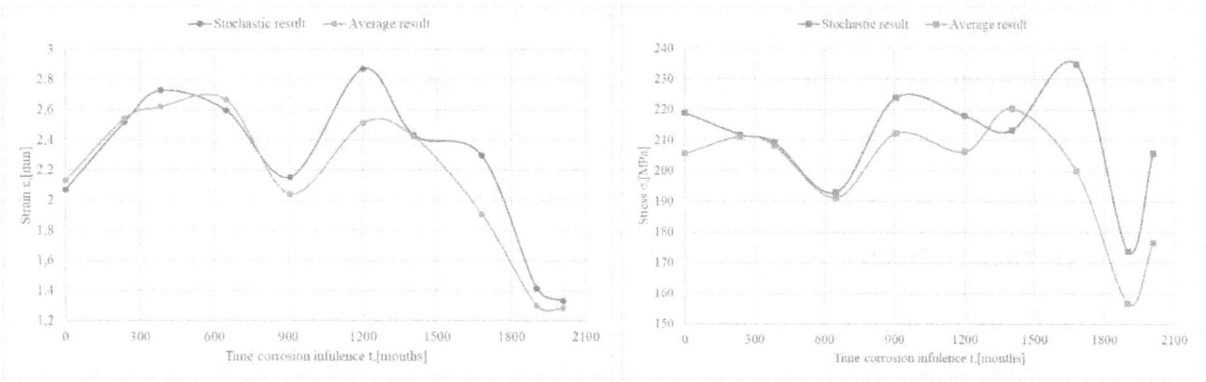

(a) (b)

Figure 3.3.1 (a) Graph of the variation of the strain deformation depending on the corrosion effect over time; [119]; (b) Graph of the change of stress strength depending on the corrosion effect over time [119];

Table 3.3.1 Result after processing [119]

	Stochastic results									
time, [months]	0	238	384	648	907	1200	1404	1680	1901	2009
ε, [mm]	2.070	2.521	2.727	2.595	2.151	2.87	2.432	2.296	1.414	1.331
σ, [MPa]	218.91	211.90	209.30	193.00	224.00	218.00	213.40	234.90	173.90	205.90
	Average results									
time, [months]	0	238	384	648	907	1200	1404	1680	1901	2009
ε, [mm]	2.130	2.543	2.619	2.664	2.039	2.511	2.423	1.907	1.299	1.282
σ, [MPa]	205.68	211.30	208.40	191.30	212.40	206.40	220.50	200.20	156.90	176.60

3.1.1. Stochastic functions

- Strain function [119]

$$\varepsilon(t) = 8.755285124854.10^{-26}.t^9 - 8.0300304061157.10^{-22}.t^8 + 3.07756854221905.10^{-18}.t^7 - 6.3859161340478285.10^{-15}.t^6 + 7.7678259285957.10^{-12}.t^5 - 5.60317265906248.10^{-9}.t^4 + 2.3101082376817885.10^{-6}.t^3 - 4.942839051062654.10^{-4}.t^2 + 4.36463169528778.10^{-2}.t + 2.070 \quad (3.3.1)$$

- Stress function [119]

$$\sigma(t) = -2.3810483526814.10^{-25}.t^9 + 4.1598070721.10^{-21}.t^8 - 2.401514963076.10^{-17}.t^7 + 6.7708808946216867.10^{-14}.t^6 - 1.0555177113064597.10^{-10}.t^5 + 5.3805413804822.10^{-8}.t^4 - 4.6105087537092795.10^{-5}.t^3 + 1.1335181228424245.10^{-2}.t^2 - 1.0891172360974972.t + 218.91 \quad (3.3.2)$$

3.2.1. Average functions

- Strain function [119]

$$\varepsilon(t) = 5.367480839453 27.10^{-26}.t^9 - 5.0922899405397.10^{-22}.t^8 + 2.02660283208.10^{-18}.t^7 - 4.3839449112770371.10^{-15}.t^6 + 5.5812486839628876.10^{-12}.t^5 - 4.229336583562.10^{-9}.t^4 + 1.837555426186442.10^{-6}.t^3 - 4.151589794458045.10^{-4}.t^2 + 3.8566723576146697.10^{-2}.t + 2.130 \quad (3.3.3)$$

- Stress function [119]

$$\sigma(t) = -2.498174109785.10^{-24}.t^9 + 2.382765260538.10^{-20}.t^8 - 9.512874598366.10^{-17}.t^7 + 2.0621927264228677.10^{-13}.t^6 - 2.6318265414139366.10^{-10}.t^5 + 2.0012980791742.10^{-7}.t^4 - 8.7125893240207485.10^{-5}.t^3 + 1.9423490246730079.10^{-2}.t^2 - 1.6587834784606574.t + 218.91 \quad (3.3.4)$$

3.2. Elastic limit point

On table 3.3.2 is present a result after processing by stochastic method and average method. In fig. 3.3.2(a) shows the relationship between the change in relative strain versus the action time of the corrosion category, and in fig. 3.3.2(b) shows the relationship between the change in strength over time due to the corrosion effect of category.

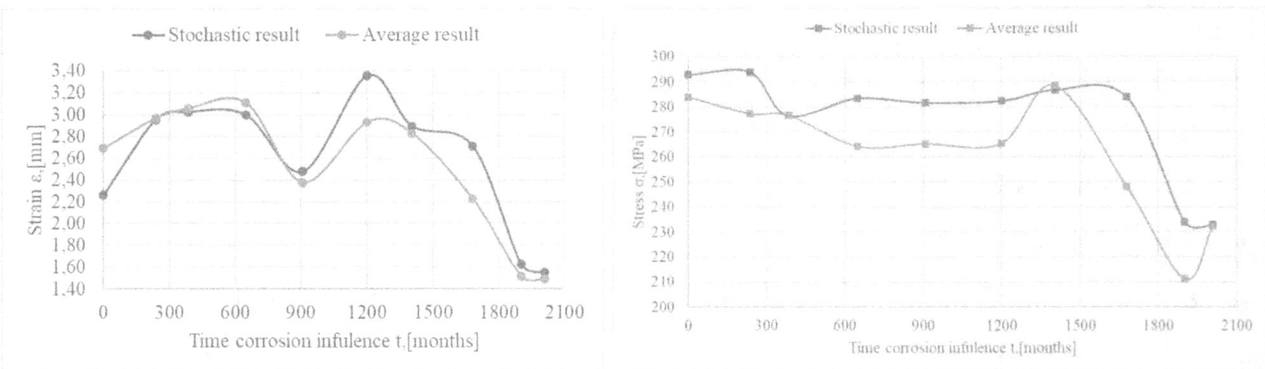

(a) (b)

Figure 3.3.2 (a) Graph of the variation of the strain deformation depending on the corrosion effect over time; (b) Graph of the change of stress strength depending on the corrosion effect over time;

Table 3.3.2 Result after processing

	Stochastic results									
time, [months]	0	238	384	648	907	1200	1404	1680	1901	2009
ε, [mm]	2.260	2.955	3.024	2.997	2.475	3.351	2.894	2.708	1.626	1.553
σ, [MPa]	292.80	293.90	276.50	283.50	281.70	282.40	286.80	283.90	233.90	232.90
	Average results									
time, [months]	0	238	384	648	907	1200	1404	1680	1901	2009
ε, [mm]	2.691	2.967	3.055	3.108	2.378	2.930	2.827	2.225	1.516	1.495
σ, [MPa]	284.00	277.30	276.60	264.10	265.00	265.10	288.50	248.00	211.20	232.20

Using the polynomial approximation [14-15,114-119] from fig. 3.3.2, functional dependence for the strain and stress point in dependence of the time of the influence of the corrosion (in months) is established in functions or equations.

3.2.1. Stochastic functions

- Strain function

$$\varepsilon\,(t) = 1.081625584.10^{-25}.t^{9} - 9.98786825896.10^{-22}.t^{8} + 3.861321553485.10^{-18}.t^{7} - 8.103193555565.10^{-15}.t^{6} + 1.0005434213068598.10^{-11}.t^{5} - 7.36511886939778.10^{-9}.t^{4} + 3.1223773618378.10^{-6}.t^{3} - 6.935354935506697.10^{-4}.t^{2} + 6.3834247657071755.10^{-2}.t + 2.260 \quad (3.3.5)$$

- Stress function

$$\sigma\,(t) = 9.70008749581398.10^{-25}.t^{9} - 9.0535136658531.10^{-21}.t^{8} + 3.6004187420368.10^{-17}.t^{7} - 7.9456831327994.10^{-14}.t^{6} + 1.0592500895146.10^{-10}.t^{5} - 8.6705020972848602.10^{-8}.t^{4} + 4.208496205445.10^{-5}.t^{3} - 1.087268957321005.10^{-2}.t^{2} + 1.092005135819282.t + 292.84 \quad (3.3.6)$$

3.2.2. Average functions

- Strain function

$$\varepsilon\,(t) = 6.24117517573.10^{-26}.t^{9} - 5.91817273278652.10^{-22}.t^{8} + 2.3536145090813.10^{-18}.t^{7} - 5.0861041382387.10^{-15}.t^{6} + 6.4649666395057758.10^{-12}.t^{5} - 4.886208899758096.10^{-9}.t^{4} \quad (3.3.7)$$

$+ 2.112736366511033.10^{-6}.t^3 - 4.723436742006428.10^{-4}.t^2 + 4.25139780806116.10^{-2}.t$

$+ 2.691$

- Stress function

$$\sigma(t) = - 2.4332378036037.10^{-24}.t^9 + 2.2384856997846.10^{-20}.t^8 - 8.60393077313.10^{-17}.t^7 + 1.7941060424665.10^{-13}.t^6 - 2.2051027618660378.10^{-10}.t^5 + 1.6228591579019729.10^{-7}.t^4 - 6.9129922037520819.10^{-5}.t^3 + 1.53854913742366.10^{-2}.t^2 - 1.3767186660402819.t + 284.00 \quad (3.3.8)$$

3.3. Yield point

On table 3.3.3 is present a result after processing by stochastic method and average method. In fig. 3.3.3(a) shows the relationship between the change in relative strain versus the action time of the corrosion category, and in fig. 3.3.3(b) shows the relationship between the change in strength over time due to the corrosion effect of category.

Table 3.3.3 Result after processing [115, 117]

	Stochastic results									
time, [months]	0	238	384	648	907	1200	1404	1680	1901	2009
ε, [mm]	5.730	5.154	6.83	6.172	3.625	4.678	4.376	3.594	2.254	1.998
σ, [MPa]	358.67	332.90	329.90	339.80	354.10	337.40	352.80	331.90	327.40	312.90
	Average results									
time, [months]	0	238	384	648	907	1200	1404	1680	1901	2009
ε, [mm]	6.010	5.547	6.664	6.363	3.863	4.459	4.285	3.358	2.065	1.921
σ, [MPa]	344.79	326.20	347.10	352.10	329.80	337.20	361.10	305.20	280.00	317.10

Using the polynomial approximation [14-15,114-119] from fig. 3.3.3, functional dependence for the strain and stress point in dependence of the time of the influence of the corrosion (in months) is established in functions or equations.

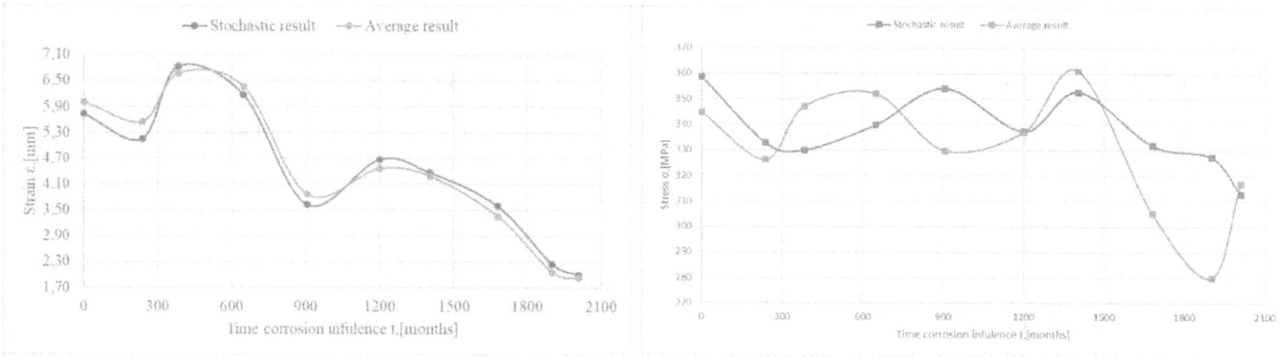

(a) (b)

Figure 3.3.3 (a) Graph of the variation of the strain deformation depending on the corrosion effect over time; [115]; (b) Graph of the change of stress strength depending on the corrosion effect over time [117];

3.3.1. Stochastic functions

- Strain function [115]

$$\varepsilon(t) = 1.3458117260815.10^{-25}.t^9 - 1.245038682557.10^{-21}.t^8 + 4.7947010266018.10^{-18}.t^7 - 9.920769341101736.10^{-15}.t^6 + 1.1855751154280902.10^{-11}.t^5 - 8.159486903279.10^{-9}.t^4 + 3.024214668908795.10^{-6}.t^3 - 5.130552906665298.10^{-4}.t^2 + 2.7100947206118678.10^{-2}.t + 5.730 \quad (3.3.9)$$

- Stress function [117]

$$\sigma(t) = -2.8792056579.10^{-24}.t^9 + 2.62353216379.10^{-20}.t^8 - 9.98395474597630 09.10^{-17}.t^7 + 2.0569398476013649.10^{-13}.t^6 - 2.48672576948199.10^{-10}.t^5 + 1.7880097435737733.10^{-7}.t^4 - 7.4001971486959507.10^{-5}.t^3 + 1.6169033086130824.10^{-2}.t^2 - 1.5173643535732102.t + 358.67 \quad (3.3.10)$$

3.3.2. Average functions

- Strain function [115]

$$\varepsilon(t) = 1.17841570377.10^{-25}.t^9 - 1.1075630018094.10^{-21}.t^8 + 4.34954814482444.10^{-18}.t^7 - 9.22557722769195.10^{-15}.t^6 + 1.13911994465885.10^{-11}.t^5 - 8.208106770310634.10^{-9}.t^4 + 3.2670105788053884.10^{-6}.t^3 - 6.2990409520477.10^{-4}.t^2 + 4.332305951862495.10^{-2}.t + 6.010 \quad (3.3.11)$$

- Stress function [117]

$$\sigma(t) = -1.447938466788.10^{-24}.t^9 + 1.269367444891.10^{-20}.t^8 - 4.62499678275497.10^{-17}.t^7 + 9.1348274101596.10^{-14}.t^6 - 1.0743412307512.10^{-10}.t^5 + 7.8123639617902388.10^{-8}.t^4 - 3.5118862703082655.10^{-5}.t^3 + 9.0871855910305294.10^{-3}.t^2 - 1.0219717267236565.t + 344.79 \quad (3.3.12)$$

3.4. Lower yield point

On table 3.3.4 is present a result after processing by stochastic method and average method. In fig. 3.3.4(a) shows the relationship between the change in relative strain versus the action time of the corrosion category, and in fig. 3.3.4(b) shows the relationship between the change in strength over time due to the corrosion effect of category.

Table 3.3.4 Result after processing [118]

	Stochastic results									
time, [months]	0	238	384	648	907	1200	1404	1680	1901	2009
ε, [mm]	8.760	9.034	9.939	9.120	5.490	6.125	5.554	5.295	2.654	2.274
σ, [MPa]	363.43	347.40	366.40	405.30	357.90	372.90	369.80	345.50	377.90	363.90
	Average results									
time, [months]	0	238	384	648	907	1200	1404	1680	1901	2009
ε, [mm]	8.950	8.366	9.836	9.174	5.008	5.569	5.339	4.173	2.399	2.132
σ, [MPa]	359.19	343.70	365.90	376.90	352.90	358.60	379.90	316.10	295.70	339.40

Using the polynomial approximation [14-15,114-119] from fig. 3.3.4, functional dependence for the strain and stress point in dependence of the time of the influence of the corrosion (in months) is established in functions or equations.

3.4.1. Stochastic functions

- Strain function [118]

$$\varepsilon(t) = 2.25155834789887.10^{-25}.t^9 - 2.0893975935.10^{-21}.t^8 + 8.11523922673499.10^{-18}.t^7 - 1.70826654789637.10^{-14}.t^6 + 2.1068180710755.10^{-11}.t^5 - 1.534413548776896.10^{-8}.t^4 + 6.31587696578745.10^{-6}.t^3 - 1.320125788994745.10^{-3}.t^2 + 1.0850319756056787.10^{-1}.t + 8.760 \quad (3.3.13)$$

- Stress function [118]

$$\sigma(t) = 2.6758181837436.10^{-24}.t^9 - 2.66556633761.10^{-20}.t^8 + 1.11051708751.10^{-16}.t^7$$
$$-2.5061812163330238.10^{-13}.t^6 + 3.31351475554.10^{-10}.t^5 - 2.5872753948088745.10^{-7}.t^4$$
$$+ 1.138941126643777.10^{-5}.t^3 - 2.5011687045203779.10^{-2}.t^2 + 2.0311564834830795.t \qquad (3.3.14)$$
$$+ 363.43$$

3.4.2. Average functions

- Strain function [118]

$$\varepsilon(t) = 1.78922874991.10^{-25}.t^9 - 1.686456365998769.10^{-21}.t^8 + 6.646539778233.10^{-18}.t^7$$
$$- 1.4162284783439 65.10^{-14}.t^6 + 1.75940192762.10^{-11}.t^5 - 1.2787397935997077.10^{-8}.t^4$$
$$+5.15841924326443.10^{-6}.t^3 - 1.019613643364765.10^{-3}.t^2 + 7.3638062870399326.10^{-2}.t \qquad (3.3.15)$$
$$+ 8.950$$

- Stress function [118]

$$\sigma(t) = -1.3780465243551.10^{-24}.t^9 + 1.18525642971.10^{-20}.t^8 - 4.21454790060257.10^{-17}.t^7 + 8.07231074792308.10^{-14}.t^6 - 9.1529600238167976.10^{-11}.t^5 +$$
$$6.414970541536.10^{-8}.t^4 - 2.8193512960264935.10^{-5}.t^3 + 7.3526410139476139.10^{-3}.t^2 \qquad (3.3.16)$$
$$- 0.843644849369.t \qquad + 359.19$$

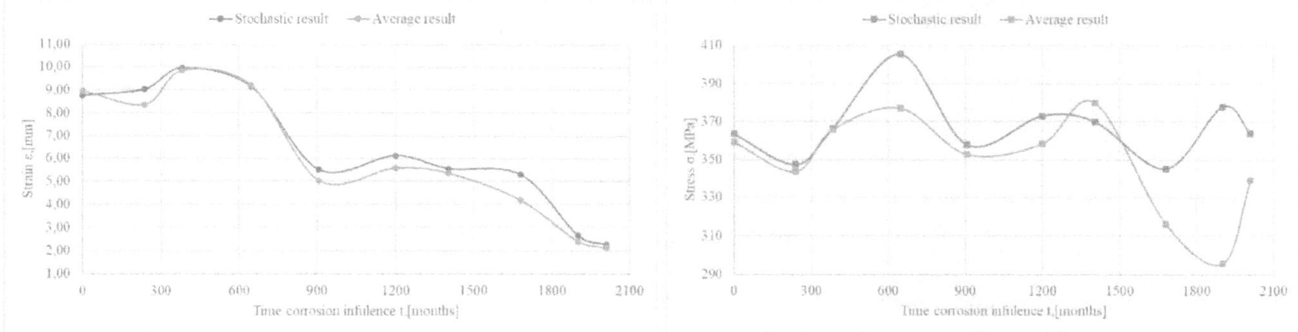

(a) (b)

Figure 3.3.4 (a) Graph of the variation of the strain deformation depending on the corrosion effect over time; [118]; (b) Graph of the change of stress strength depending on the corrosion effect over time [118];

3.5.Ultimate strength point

On table 3.3.5 is present a result after processing by stochastic method and average method. In fig. 3.3.5(a) shows the relationship between the change in relative strain versus the action time of the corrosion category, and in fig. 3.3.5(b) shows the relationship between the change in strength over time due to the corrosion effect of category.

Table 3.3.5 Result after processing [114]

	Stochastic results									
time, [months]	0	238	384	648	907	1200	1404	1680	1901	2009
ε, [mm]	28.540	28.370	23.790	19.730	15.270	13.120	11.250	8.090	6.070	3.430
σ, [MPa]	460.91	448.90	450.90	474.80	444.00	446.90	449.90	413.90	433.90	451.90
	Average results									
time, [months]	0	238	384	648	907	1200	1404	1680	1901	2009
ε, [mm]	29.600	23.410	23.440	19.440	12.470	12.250	12.000	8.890	4.550	3.550
σ, [MPa]	447.20	437.10	441.30	444.20	423.70	431.40	443.00	390.60	354.40	421.00

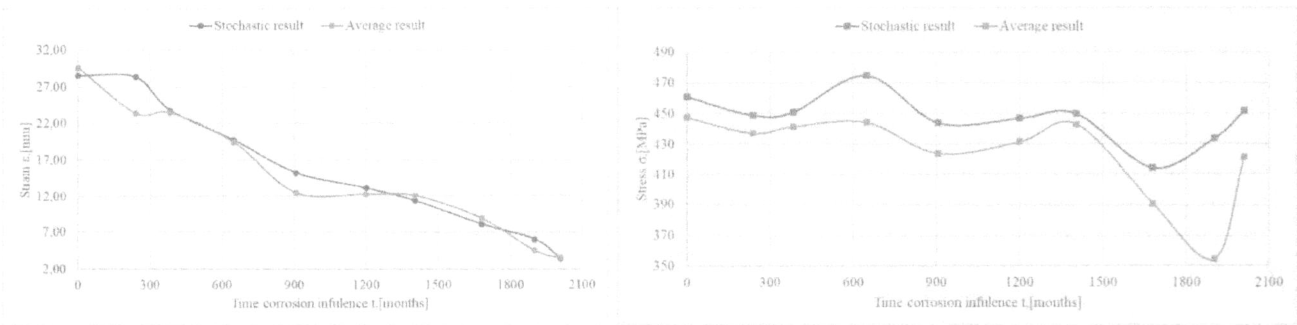

(a) (b)

Figure 3.3.5 (a) Graph of the variation of the strain deformation depending on the corrosion effect over time; [114]; (b) Graph of the change of stress strength depending on the corrosion effect over time [114];

Using the polynomial approximation [14-15,114-119] from fig. 3.3.5, functional dependence for the strain and stress point in dependence of the time of the influence of the corrosion (in months) is established in functions or equations.

3.5.1. Stochastic functions

- Strain function [114]

$$\varepsilon(t) = 1.195341055488.10^{-25}.t^9 - 1.1924766435298.10^{-21}.t^8 + 5.0108867563188.10^{-20}.t^7 - 1.15265155854.10^{-14}.t^6 + 1.57876482878837.10^{-11}.t^5 - 1.3114512660164491.10^{-8}.t^4 + 6.430856848762359.10^{-6}.t^3 - 1.70442843909736.10^{-3}.t^2 + 1.74756978332975.10^{-1}.t + 28.540 \quad (3.3.17)$$

- Stress function [114]

$$\sigma(t) = 8.479164674075.10^{-25}.t^9 - 9.373202973409837.10^{-21}.t^8 + 4.2860332208922.10^{-17}.t^7 - 1.052018526866775.10^{-13}.t^6 + 1.501616838599.10^{-10}.t^5 - 1.2587365611160668.10^{-7}.t^4 + 5.929032463175279.10^{-5}.t^3 - 1.3928175993827889.10^{-2}.t^2 + 1.194082923412319.t + 460.91 \quad (3.3.18)$$

3.5.2. Average functions

- Strain function [114]

$$\varepsilon(t) = 1.71143994825.10^{-25}.t^9 - 1.61772955395387.10^{-21}.t^8 + 6.384690411036.10^{-18}.t^7 - 1.3577473689315828.10^{-14}.t^6 + 1.671863240849.10^{-11}.t^5 - 1.1882383914570681.10^{-8}.t^4 + 4.5564955529305888.10^{-6}.t^3 - 7.91351198781722.10^{-4}.t^2 + 2.0058431053142894^{-1}.t + 29.600 \quad (3.3.19)$$

- Stress function [114]

$$\sigma(t) = 3.461893905814118.10^{-26}.t^9 - 7.68952611017.10^{-22}.t^8 + 5.2099057095328.10^{-18}.t^7 - 1.64126515576645.10^{-14}.t^6 + 2.7365758559312739.10^{-11}.t^5 - 2.48096922823416.10^{-8}.t^4 + 1.168794061096096.10^{-5}.t^3 - 2.4765148030392275.10^{-3}.t^2 + 0.143203205276329.t + 460.91 \quad (3.3.20)$$

3.6. Fracture point

On table 3.3.6 is present a result after processing by stochastic method and average method. In fig. 3.3.6(a) shows the relationship between the change in relative strain versus the action time of the corrosion category, and in fig. 3.3.6(b) shows the relationship between the change in strength over time due to the corrosion effect of category.

Using the polynomial approximation [14-15,114-119] from fig. 3.3.6, functional dependence for the strain and stress point in dependence of the time of the influence of the corrosion (in months) is established in functions or equations.

Table 3.3.6 Result after processing

	Stochastic results									
time, [months]	0	238	384	648	907	1200	1404	1680	1901	2009
ε, [mm]	40.410	36.080	28.360	23.420	17.600	16.640	12.990	10.230	6.750	5.170
σ, [MPa]	286.41	282.70	284.40	297.40	288.90	285.90	305.90	289.90	287.90	322.90
	Average results									
time, [months]	0	238	384	648	907	1200	1404	1680	1901	2009
ε, [mm]	38.800	29.180	28.250	23.020	15.120	14.370	14.160	10.280	5.630	4.310
σ, [MPa]	285.08	284.90	290.50	304.90	292.70	298.30	300.30	263.70	231.00	295.30

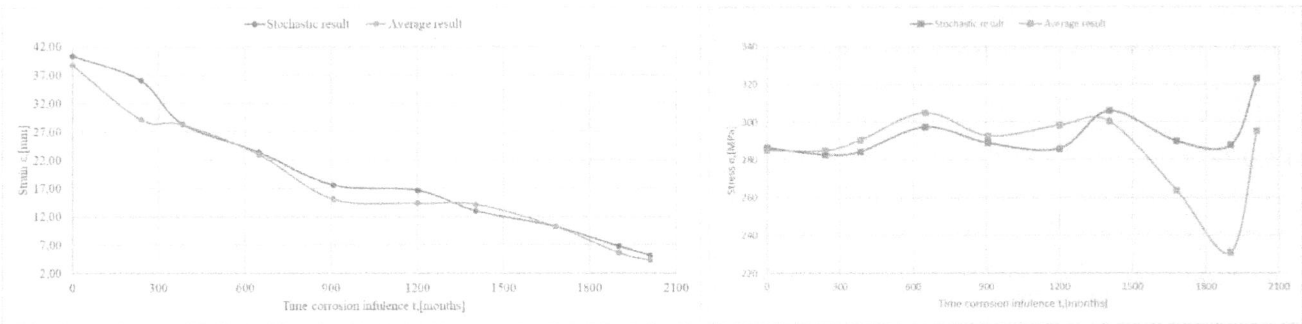

(a) (b)

Figure 3.3.6 (a) Graph of the variation of the strain deformation depending on the corrosion effect over time; (b) Graph of the change of stress strength depending on the corrosion effect over time;

3.6.1. Stochastic functions

- Strain function

$$\varepsilon (t) = 4.3244785547148.10^{-25}.t^{9} - 4.096292319.10^{-21}.t^{8} + 1.6327387456136.10^{-17}.t^{7} - 3.55780169017.10^{-14}.t^{6} + 4.607468769365479.10^{-11}.t^{5} - 3.60721194794395.10^{-8}.t^{4} + 1.6558571540541856.10^{-6}.t^{3} - 4.0376517069226.10^{-3}.t^{2} + 3.6766377927091365.10^{-1}.t + 40.410 \quad (3.3.21)$$

- Stress function

$$\sigma (t) = -9.448642244401.10^{-25}.t^{9} + 8.15143809294.10^{-21}.t^{8} - 2.8756561293233.10^{-17}.t^{7} +5.3426373520476.10^{-14}.t^{6} - 5.6172627135702045.10^{-11}.t^{5} +3.3548887913095673.10^{-8}.t^{4} \quad (3.3.22)$$

$- 1.0985833549051269.10^{-5}.t^3 + 1.878287267987464.10^{-3}.t^2 - 1.4830380440889224.t + 286.41$

3.6.2. Average functions

- Strain function

$$\varepsilon (t) = 1.4367762759987.10^{-25}.t^9 - 1.37489147763.10^{-21}.t^8 + 5.4885038098994.10^{-18}.t^7 - 1.178133150394276.10^{-14}.t^6 + 1.4583187831830225.10^{-11}.t^5 - 1.0332266570377.10^{-8}.t^4 + 3.8708476039739035.10^{-6}.t^3 - 6.08069651991294.10^{-4}.t^2 - 1.440134604417975.10^{-2}.t + 38.800 \quad (3.3.23)$$

- Stress function

$$\sigma (t) = 8.667467178089.10^{-25}.t^9 - 8.2296402604764.10^{-21}.t^8 + 3.319245554101767.10^{-17}.t^7 - 7.353937405780503.10^{-14}.t^6 + 9.65794463566538.10^{-11}.t^5 - 7.563524764287225.10^{-8}.t^4 + 3.3683678422741049.10^{-5}.t^3 - 7.5630338599963695.10^{-3}.t^2 + 6.5151986221534175.t + 285.08 \quad (3.3.24)$$

3.7. Young's modulus (E-modulus)

On table 3.3.7 is present a result is presented after processing, as the given values of the E-modulus are calculated as the tangent of the angle that the deformation curve makes with the abscissa axis of the σ-ε diagram and subsequently (the obtained values) are processed using the stochastic method and average method. Fig. 3.3.7 shows the relationship between the change in E-modulus and the time of action of the corrosion category.

Table 3.3.7 Result after processing

	Stochastic results									
time, [months]	0	238	384	648	907	1200	1404	1680	1901	2009
E, [GPa]	211.50	168.10	153.50	148.70	156.20	151.90	131.60	153.50	184.50	232.00
	Average results									
time, [months]	0	238	384	648	907	1200	1404	1680	1901	2009
E, [GPa]	193.10	166.20	159.10	143.60	156.30	164.40	136.50	157.50	181.20	206.60

Using the polynomial approximation [14-15,114-119] from fig. 3.3.7, functional dependence for the E-modulus in dependence of the time of the influence of the corrosion (in months) is established in functions or equations.

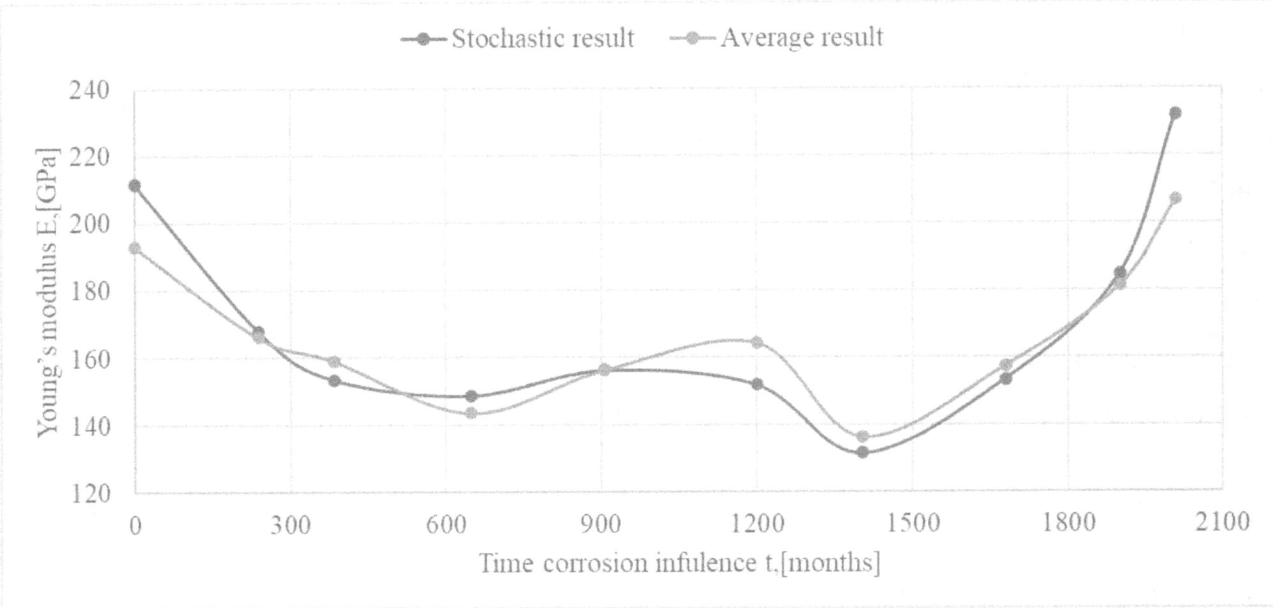

Figure 3.3.7 Graph of the variation of the Young's modulus (E-modulus) depending on the corrosion effect over time;

3.7.1. Stochastic functions

$$E\,(t) = 1.3650061233844.10^{-24}.t^9 - 1.2027136376638.10^{-20}.t^8 + 4.41480460652.10^{-17}.t^7 - 8.764108563220423.10^{-14}.t^6 + 1.0232326240667461.10^{-10}.t^5 - 7.1551894138704137.10^{-8}.t^4 + 2.896930460937739.10^{-5}.t^3 - 5.918809202864088.10^{-3}.t^2 + 0.28109751154078.t + 211.50 \quad (3.3.25)$$

3.7.2. Average functions

$$E\,(t) = 1.7707300877.10^{-24}.t^9 - 1.54141285266356.10^{-20}.t^8 + 5.5380749098601536.10^{-17}.t^7 - 1.061038971755.10^{-13}.t^6 + 1.1705331118482578.10^{-10}.t^5 - 7.5060551228578404.10^{-8}.t^4 + 2.690249323494332.10^{-5}.t^3 - 4.7603603834720735.10^{-3}.t^2 + 0.20402599763492.t + 193.10 \quad (3.3.26)$$

4. Corrosion category C4

Standard ISO 12944-5 has defined the existence of this category, in coastal climatic and industrial zone with moderate salinity, chemical industry, shipbuilding and vessels.

4.1. Point of the proportional limit

On table 3.4.1 is present a result after processing by stochastic method and average method. In fig. 3.4.1(a) shows the relationship between the change in relative strain versus the action time of the corrosion category, and in fig. 3.4.1(b) shows the relationship between the change in strength over time due to the corrosion effect of category.

Table 3.4.1 Result after processing [119]

	Stochastic results									
time, [months]	0	149	240	405	567	750	878	1050	1188	1256
ε, [mm]	2.070	2.521	2.727	2.595	2.151	2.87	2.432	2.296	1.414	1.331
σ, [MPa]	218.91	211.90	209.30	193.00	224.00	218.00	213.40	234.90	173.90	205.90
	Average results									
time, [months]	0	149	240	405	567	750	878	1050	1188	1256
ε, [mm]	2.130	2.543	2.619	2.664	2.039	2.511	2.423	1.907	1.299	1.282
σ, [MPa]	205.68	211.30	208.40	191.30	212.40	206.40	220.50	200.20	156.90	176.60

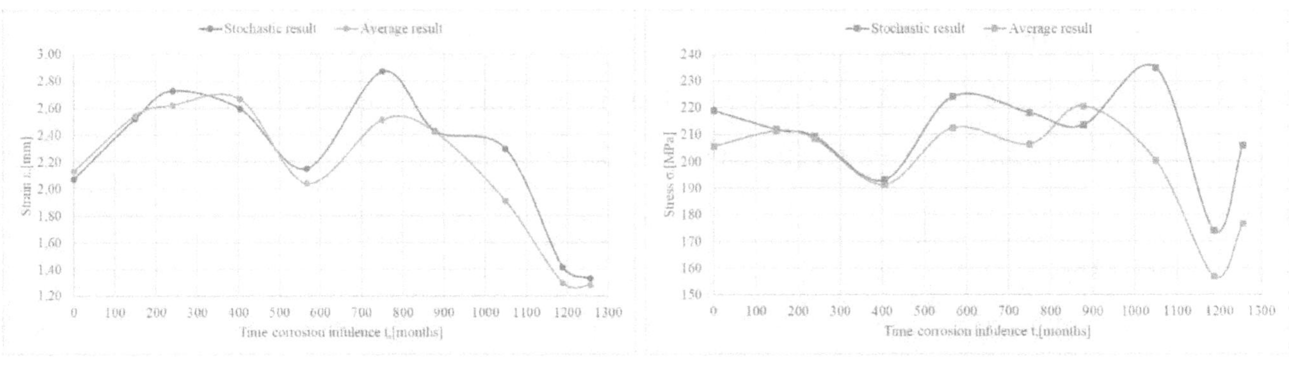

(a) (b)

Figure 3.4.1 (a) Graph of the variation of the strain deformation depending on the corrosion effect over time; [119]; (b) Graph of the change of stress strength depending on the corrosion effect over time [119];

Using the polynomial approximation [14-15,114-119] from fig. 3.4.1, functional dependence for the strain and stress point in dependence of the time of the influence of the corrosion (in months) is established in functions or equations.

4.1.1. Stochastic functions

- Strain function [119]

$$\varepsilon(t) = 5.999835992033.10^{-24}.t^9 - 3.4397511034137.10^{-20}.t^8 + 8.2406463399957.10^{-17}.t^7 - 1.0688667603050666.10^{-13}.t^6 + 8.1273655567123.10^{-11}.t^5 - 3.6646980822792728.10^{-8}.t^4 + 9.4449313640337727.10^{-6}.t^3 - 1.2633669929962147.10^{-3}.t^2 + 6.975213399079.10^{-2}.t + 2.070 \quad (3.4.1)$$

- Stress function [119]

$$\sigma(t) = -1.62302845914.10^{-23}.t^9 + 1.778346159187.10^{-19}.t^8 - 6.425750922648.10^{-16}.t^7 + 1.1332601713862929.10^{-12}.t^6 - 1.10482673965.10^{-9}.t^5 + 6.139816502575666.10^{-7}.t^4 - 1.8869617308085507.10^{-4}.t^3 + 2.9010198373649605.10^{-2}.t^2 - 1.743272971564717.t + 218.91 \quad (3.4.2)$$

4.1.2. Average functions

- Strain function [119]

$$\varepsilon(t) = 3.677737129237.10^{-24}.t^9 - 2.181255890899.10^{-20}.t^8 + 5.4268816867337.10^{-17}.t^7 - 7.33905460685376.10^{-14}.t^6 + 85.841247968096.10^{-11}.t^5 - 2.7672888949467169.10^{-8}.t^4 + 7.51704846841126.10^{-6}.t^3 - 1.0618942801698119.10^{-3}.t^2 + 6.168864517046278.10^{-2}.t + 2.130 \quad (3.4.3)$$

- Stress function [119]

$$\sigma(t) = -1.71351808686.10^{-22}.t^9 + 1.021631416362.10^{-18}.t^8 - 2.549649100602.10^{-15}.t^7 + 3.45512908488927.10^{-12}.t^6 - 2.7565913560054436.10^{-9}.t^5 + 1.310465889617915.10^{-6}.t^4 - 3.5668806282743891.10^{-4}.t^3 + 4.9721724169531696.10^{-2}.t^2 - 2.6557315385659646.t + 205.68 \quad (3.4.4)$$

4.2. Elastic limit point

On table 3.4.2 is present a result after processing by stochastic method and average method. In fig. 3.4.2(a) shows the relationship between the change in relative strain versus the action time of the corrosion category, and in fig. 3.4.2(b) shows the relationship between the change in strength over time due to the corrosion effect of category.

Table 3.4.2 Result after processing

	Stochastic results									
time, [months]	0	149	240	405	567	750	878	1050	1188	1256
ε, [mm]	2.260	2.955	3.024	2.997	2.475	3.351	2.894	2.708	1.626	1.553
σ, [MPa]	292.80	293.90	276.50	283.50	281.70	282.40	286.80	283.90	233.90	232.90
	Average results									
time, [months]	0	149	240	405	567	750	878	1050	1188	1256
ε, [mm]	2.691	2.967	3.055	3.108	2.378	2.930	2.827	2.225	1.516	1.495
σ, [MPa]	284.00	277.30	276.60	264.10	265.00	265.10	288.50	248.00	211.20	232.20

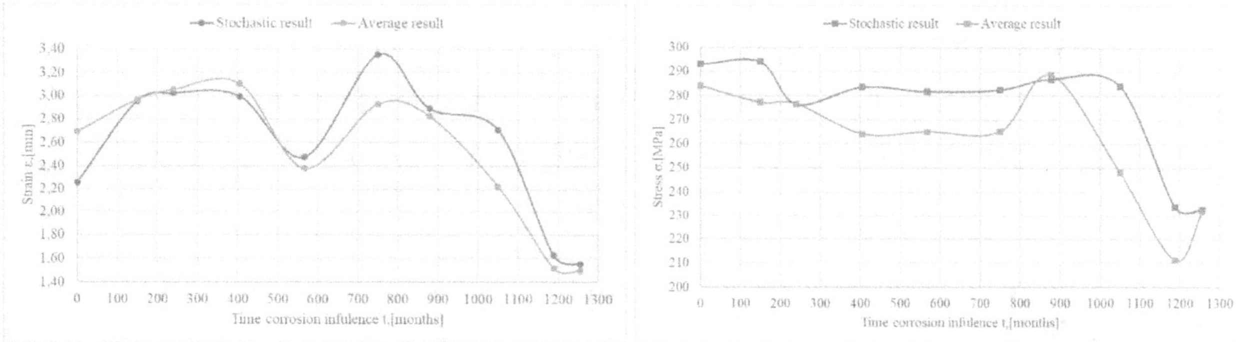

(a) (b)

Figure 3.4.2 (a) Graph of the variation of the strain deformation depending on the corrosion effect over time; (b) Graph of the change of stress strength depending on the corrosion effect over time;

Using the polynomial approximation [14-15,114-119] from fig. 3.4.2, functional dependence for the strain and stress point in dependence of the time of the influence of the corrosion (in months) is established in functions or equations.

4.2.1. Stochastic functions

- Strain function

$$\varepsilon(t) = 7.4148753196472.10^{-24}.t^{9} - 4.280084387196.10^{-20}.t^{8} + 1.0343657845307.10^{-16}.t^{7} -1.3569397611993469.10^{-13}.t^{6} + 1.0474087274049783.10^{-10}.t^{5} - 4.820029970469.10^{-8}.t^{4} + 1.2775202879245376.10^{-5}.t^{3} - 1.774207660138187.10^{-3}.t^{2} + 0.10211964195281378.t + 2.260 \quad (3.4.5)$$

- Stress function

$$\sigma(t) = 6.68512645585.10^{-23}.t^9 - 3.900431817.10^{-19}.t^8 + 9.6962057302697078.10^{-16}.t^7 - 1.3375983192967.10^{-12}.t^6 + 1.1146409618654173.10^{-9}.t^5 - 5.7033181472587681.10^{-7}.t^4 + 1.730520515465055.10^{-4}.t^3 - 2.795122902670239.10^{-2}.t^2 + 1.755670020652365.t + 292.84 \quad (3.4.6)$$

4.2.2. Average functions

- Strain function

$$\varepsilon(t) = 4.2763985278608.10^{-24}.t^9 - 1\ 2.535028247842.10^{-20}.t^8 + 6.302586398636.10^{-17}.t^7 - 8.5145597577605002.10^{-14}.t^6 + 6.766167048599.10^{-11}.t^5 - 3.1971045255217644.10^{-8}.t^4 + 8.64282073985516.10^{-6}.t^3 - 1.208175557301615.10^{-3}.t^2 + 6.8004217706800429.10^{-2}.t + 2.691 \quad (3.4.7)$$

- Stress function

$$\sigma(t) = -1.669520084362.10^{-22}.t^9 + 9.6002634214636.10^{-19}.t^8 - 2.306518126472.10^{-15}.t^7 + 3.0064360900523688.10^{-12}.t^6 - 2.309903916055907.10^{-9}.t^5 + 1.06274916230838.10^{-6}.t^4 - 2.8303116203732875.10^{-4}.t^3 + 3.9386583154122898.10^{-2}.t^2 - 2.204076658908666.t + 284.00 \quad (3.4.8)$$

4.3. Yield point

On table 3.4.3 is present a result after processing by stochastic method and average method. In fig. 3.4.3(a) shows the relationship between the change in relative strain versus the action time of the corrosion category, and in fig. 3.4.3(b) shows the relationship between the change in strength over time due to the corrosion effect of category.

Using the polynomial approximation [14-15,114-119] from fig. 3.4.3, functional dependence for the strain and stress point in dependence of the time of the influence of the corrosion (in months) is established in functions or equations.

4.3.1. Stochastic functions

- Strain function [115]

$$\varepsilon(t) = 9.212123520033.10^{-24}.t^9 - 5.3267905281249.10^{-20}.t^8 + 1.28214570400958.10^{-17}.t^7 - 1.658016648091.10^{-13}.t^6 + 1.2382074136909.10^{-11}.t^5 - 5.3242177108261268.10^{-8}.t^4 + 1.232345060581979.10^{-6}.t^3 - 1.3040287631467258.10^{-3}.t^2 + 4.279285991344129.10^{-2}.t + 5.730 \quad (3.4.9)$$

- Stress function [117]

$$\sigma (t) = - 1.9736542324201.10^{-22}.t^9 + 1.12416035976344.10^{-18}.t^8 - 2.674171223.10^{-15}.t^7 + 3.443937833181775.10^{-12}.t^6 - 2.6026322654754967.10^{-9}.t^5 + 1.1698051487062.10^{-6}.t^4 - 3.026645239841166.10^{-4}.t^3 + 4.134404776219415.10^{-2}.t^2 - 2.4260671844246526.t + 358.67 \quad (3.4.10)$$

4.3.2. Average functions

- Strain function [115]

$$\varepsilon (t) = 8.0677669742382.10^{-24}.t^9 - 4.73985338866766.10^{-20}.t^8 + 1.1635284931.10^{-17}.t^7 - 1.54259578449202.10^{-13}.t^6 + 1.190510860045798.10^{-10}.t^5 - 5.36134826134507.10^{-8}.t^4 + 1.33344579353654.10^{-5}.t^3 - 1.6060088247672349.10^{-3}.t^2 + 6.8952904582559441.10^{-2}.t + 6.010 \quad (3.4.11)$$

- Stress function [117]

$$\sigma (t) = - 9.9687895642.10^{-23}.t^9 + 5.46301924310069.10^{-19}.t^8 - 1.2443631144538.10^{-15}.t^7 + 1.5366394309015498.10^{-12}.t^6 - 1.1300500322995734.10^{-9}.t^5 + 5.13868021382112.10^{-7}.t^4 - 1.44444112791403.10^{-4}.t^3 + 2.3366296307413464.10^{-2}.t^2 - 1.6425061743405691.t + 344.79 \quad (3.4.12)$$

Table 3.4.3 Result after processing [115, 117]

	Stochastic results									
time, [months]	0	149	240	405	567	750	878	1050	1188	1256
ε, [mm]	5.730	5.154	6.830	6.172	3.625	4.678	4.376	3.594	2.254	1.998
σ, [MPa]	358.67	332.90	329.90	339.80	354.10	337.40	352.80	331.90	327.40	312.90
	Average results									
time, [months]	0	149	240	405	567	750	878	1050	1188	1256
ε, [mm]	6.010	5.547	6.664	6.363	3.863	4.459	4.285	3.358	2.065	1.921
σ, [MPa]	344.79	326.20	347.10	352.10	329.80	337.20	361.10	305.20	280.00	317.10

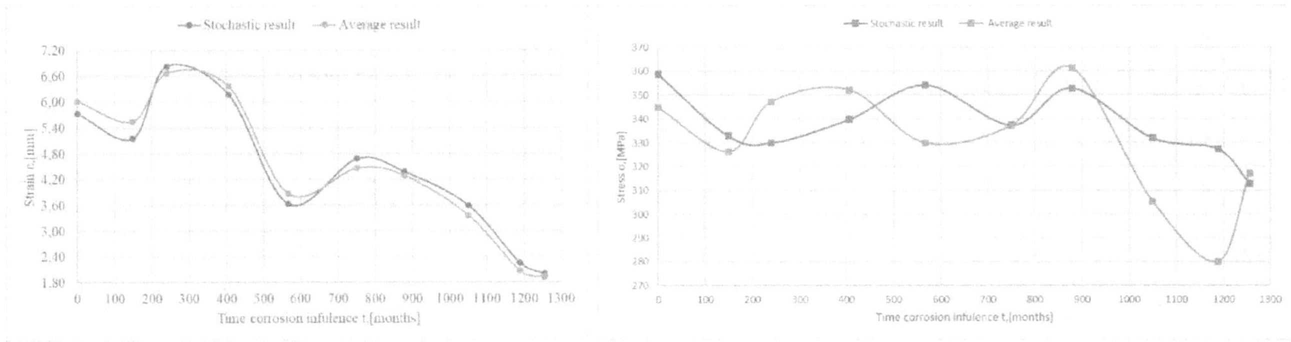

(a) (b)

Figure 3.4.3 (a) Graph of the variation of the strain deformation depending on the corrosion effect over time; [115]; (b) Graph of the change of stress strength depending on the corrosion effect over time [117];

4.4.Lower yield point

On table 3.4.4 is present a result after processing by stochastic method and average method. In fig. 3.4.4(a) shows the relationship between the change in relative strain versus the action time of the corrosion category, and in fig. 3.4.4(b) shows the relationship between the change in strength over time due to the corrosion effect of category.

Table 3.4.4 Result after processing [118]

	Stochastic results									
time, [months]	0	149	240	405	567	750	878	1050	1188	1256
ε, [mm]	8.760	9.034	9.939	9.120	5.490	6.125	5.554	5.295	2.654	2.274
σ, [MPa]	363.43	347.40	366.40	405.30	357.90	372.90	369.80	345.50	377.90	363.90
	Average results									
time, [months]	0	149	240	405	567	750	878	1050	1188	1256
ε, [mm]	8.950	8.366	9.836	9.174	5.008	5.569	5.339	4.173	2.399	2.132
σ, [MPa]	359.19	343.70	365.90	376.90	352.90	358.60	379.90	316.10	295.70	339.40

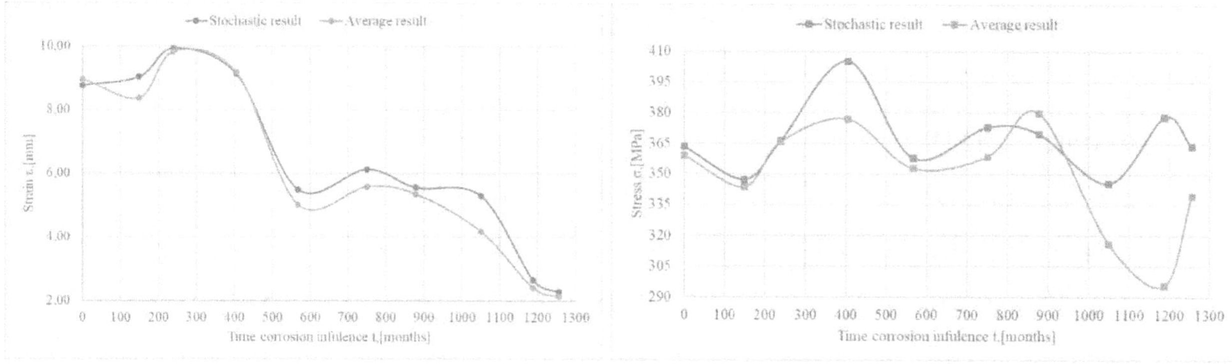

(a) (b)

Figure 3.4.4 (a) Graph of the variation of the strain deformation depending on the corrosion effect over time; [118]; (b) Graph of the change of stress strength depending on the corrosion effect over time [118];

Using the polynomial approximation [14-15,114-119] from fig. 3.4.4, functional dependence for the strain and stress point in dependence of the time of the influence of the corrosion (in months) is established in functions or equations.

4.4.1. Stochastic functions

- Strain function [118]

$$\varepsilon(t) = 1.543301832192.10^{-23}.t^{9} - 8.951922075812.10^{-20}.t^{8} + 2.1733287435017.10^{-16}.t^{7} - 2.859627026274.10^{-13}.t^{6} + 2.2044988654785645.10^{-10}.t^{5} - 1.0035797215693892.10^{-7}.t^{4} + 2.5820600508053.10^{-5}.t^{3} - 3.373462096500475.10^{-3}.t^{2} + 1.7332965212903578.10^{-1}.t + 8.760 \tag{3.4.13}$$

- Stress function [118]

$$\sigma(t) = 1.832605124196.10^{-22}.t^{9} - 1.1413775472924.10^{-18}.t^{8} + 2.972958532406.10^{-15}.t^{7} - 4.19465307430389.10^{-12}.t^{6} + 3.4672793745262669.10^{-9}.t^{5} - 1.6926079513134389.10^{-6}.t^{4} + 4.6583273544976186.10^{-4}.t^{3} - 6.3958754402950874.10^{-2}.t^{2} + 3.24775325193426.t + 363.43 \tag{3.4.14}$$

4.4.2. Average functions

- Strain function [118]

$$\varepsilon(t) = 1.225213551045.10^{-23}.t^{9} - 7.21876416681574.10^{-20}.t^{8} + 1.778368631884.10^{-16}.t^{7} - 2.36859257071396.10^{-13}.t^{6} + 1.8392383911633.10^{-10}.t^{5} - 8.3549349578466564.10^{-8}.t^{4} \tag{3.4.15}$$

$+ 2.106267744777402.10^{-5}.t^3 - 2.6012128580769388.10^{-3}.t^2 + 1.17336350868907.10^{-1}.t + 8.950$

- Stress function [118]

$$\sigma(t) = -9.50462004085.10^{-23}.t^9 + 5.11069402493.10^{-19}.t^8 - 1.13624555698744.10^{-15}.t^7 +1.3609505088257337.10^{-12}.t^6 -9.6515718975798.10^{-10}.t^5+4.2310081389470598.10^{-7} t^4 - 1.1627795433348605.10^{-4}.t^3 + 1.895098359574536.10^{-2}.t^2 - 1.358379964545405.t + 359.19 \quad (3.4.16)$$

4.5. Ultimate strength point

On table 3.4.5 is present a result after processing by stochastic method and average method. In fig. 3.4.5(a) shows the relationship between the change in relative strain versus the action time of the corrosion category, and in fig. 3.4.5(b) shows the relationship between the change in strength over time due to the corrosion effect of category.

Using the polynomial approximation [14-15,114-119] from fig. 3.4.5, functional dependence for the strain and stress point in dependence of the time of the influence of the corrosion (in months) is established in functions or equations.

4.5.1. Stochastic functions

- Strain function [114]

$$\varepsilon(t) = 8.1703438301396.10^{-24}.t^9 - 5.09795767501378.10^{-20}.t^8 + 1.33993403853.10^{-16}.t^7 - 1.92804367004616.10^{-13}.t^6 + 1.65202930815434.10^{-10}.t^5 - 8.5856013077865.10^{-8}.t^4 + 2.634186662441486.10^{-5}.t^3 - 4.36866115084418.10^{-3}.t^2 + 2.80329725594578.10^{-1}.t + 28.540 \quad (3.4.17)$$

- Stress function [114]

$$\sigma(t)= 5.78020991405889.10^{-23}.t^9 - 4.0001611209649.10^{-19}.t^8 + 1.1447038899712.10^{-15}.t^7 - 1.75792592973828.10^{-12}.t^6 + 1.5696314522215.10^{-9}.t^5 - 8.2296472530107166.10^{-7}.t^4 + 2.424427350933963.10^{-4}.t^3 - 3.56209007721412.10^{-2}.t^2 + 1.9103328891199964.t + 460.91 \quad (3.4.18)$$

4.5.2. Average functions

- Strain function [114]

$$\varepsilon(t) = 1.1707609783577.10^{-23}.t^9 - 6.918017923998.10^{-20}.t^8 + 1.706810053937.10^{-16}.t^7 - 12.26896981841396.10^{-13}.t^6 + 1.74646743817528.10^{-10}.t^5 - 7.7585981422743.10^{-8}.t^4 \quad (3.4.19)$$

$$+ 1.859387994044886.10^{-5}.t^{3} - 2.017526756017038.10^{-3}.t^{2} + 3.1740649703995.10^{-2}.t + 29.600$$

- Stress function [114]

$$\sigma (t) = 1.8584602337.10^{-24}.t^{9} - 3.022360441977347.10^{-20}.t^{8} + 1.3352727341714.10^{-16}.t^{7} - 2.67595756979779.10^{-13}.t^{6} + 2.8133139223656856.10^{-10}.t^{5} - 1.6016226079085234.10^{-7}.t^{4} + 4.726863353378447.10^{-5}.t^{3} - 6.2618447794189468.10^{-3}.t^{2} + 0.225203582432704.t + 447.20 \quad (3.4.20)$$

Table 3.4.5 Result after processing [114]

	Stochastic results									
time, [months]	0	149	240	405	567	750	878	1050	1188	1256
ε, [mm]	28.540	28.370	23.790	19.730	15.270	13.120	11.250	8.090	6.070	3.430
σ, [MPa]	460.91	448.90	450.90	474.80	444.00	446.90	449.90	413.90	433.90	451.90
	Average results									
time, [months]	0	149	240	405	567	750	878	1050	1188	1256
ε, [mm]	29.600	23.410	23.440	19.440	12.470	12.250	12.00	8.890	4.550	3.550
σ, [MPa]	447.20	437.10	441.30	444.20	423.70	431.40	443.00	390.60	354.40	421.00

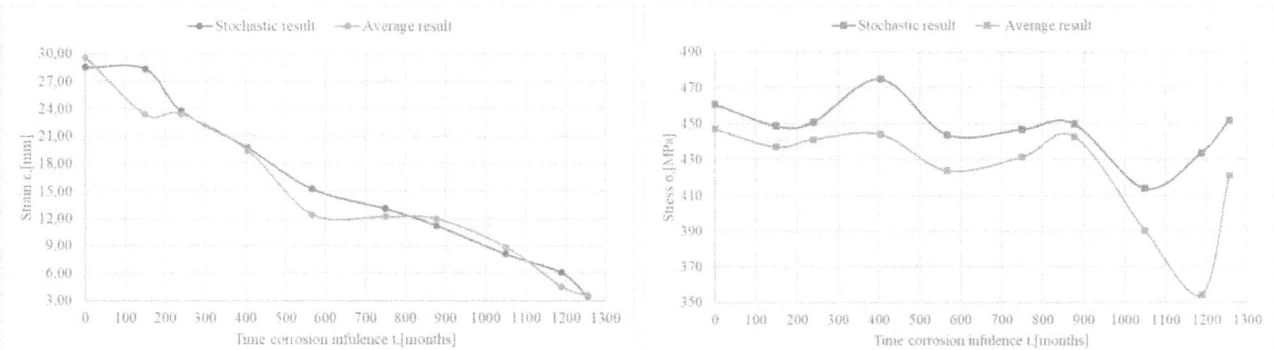

(a) (b)

Figure 3.4.5 (a) Graph of the variation of the strain deformation depending on the corrosion effect over time; [114]; (b) Graph of the change of stress strength depending on the corrosion effect over time [114];

4.6.Fracture point

On table 3.4.6 is present a result after processing by stochastic method and average method. In fig. 3.4.6(a) shows the relationship between the change in relative strain versus the action time of the corrosion category, and in fig. 3.4.6(b) shows the relationship between the change in strength over time due to the corrosion effect of category.

Table 3.4.6 Result after processing

	Stochastic results									
time, [months]	0	149	240	405	567	750	878	1050	1188	1256
ε, [mm]	40.410	36.080	28.360	23.420	17.600	16.640	12.990	10.230	6.750	5.170
σ, [MPa]	286.41	282.70	284.40	297.40	288.90	285.90	305.90	289.90	287.90	322.90
	Average results									
time, [months]	0	149	240	405	567	750	878	1050	1188	1256
ε, [mm]	38.800	29.180	28.250	23.020	15.120	14.370	14.160	10.280	5.630	4.310
σ, [MPa]	285.08	284.90	290.50	304.90	292.70	298.30	300.30	263.70	231.00	295.30

Using the polynomial approximation [14-15,114-119] from fig. 3.4.6, functional dependence for the strain and stress point in dependence of the time of the influence of the corrosion (in months) is established in functions or equations.

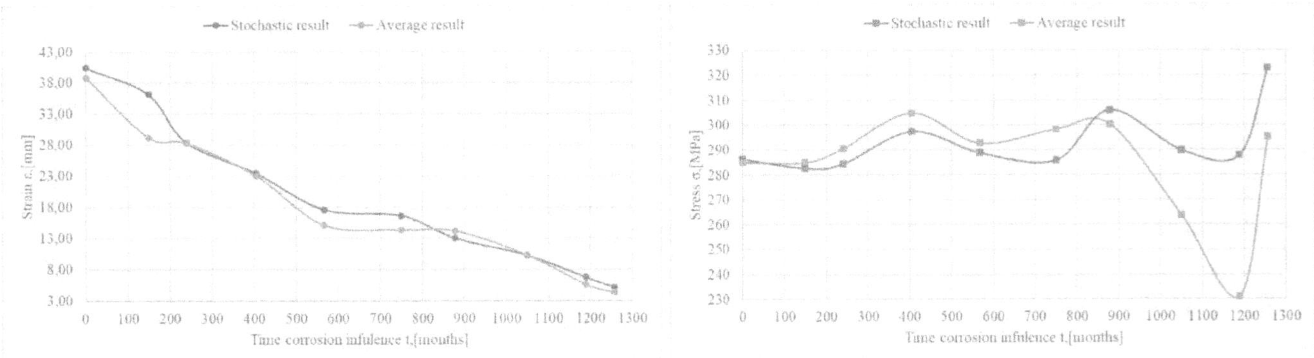

(a) (b)

Figure 3.4.6 (a) Graph of the variation of the strain deformation depending on the corrosion effect over time; (b) Graph of the change of stress strength depending on the corrosion effect over time;

4.6.1. Stochastic functions

- Strain function

$$\varepsilon(t) = 2.96322199430685.10^{-23}.t^9 - 1.75492621781.10^{-19}.t^8 + 4.3736660521274.10^{-16}.t^7 - 5.9593762050496477.10^{-13}.t^6 + 4.82623135079735.10^{-10}.t^5 - 2.363146007065.10^{-7}.t^4 + 6.78538544375019.10^{-5}.t^3 - 1.03511710391633.10^{-2}.t^2 + 5.8993890145587136.10^{-1}.t + 40.410 \quad (3.4.21)$$

- Stress function

$$\sigma(t) = -6.477745199172.10^{-23}.t^9 + 3.492182306405.10^{-19}.t^8 - 7.6982881816139.10^{-16}.t^7 + 8.93678410248044.10^{-13}.t^6 - 5.870428642333748.10^{-10}.t^5 + 2.1900495574567698.10^{-7}.t^4 - 4.4781365985443778.10^{-5}.t^3 + 4.7798734544418076.10^{-3}.t^2 - 2.35784195325717.t + 286.41 \quad (3.4.22)$$

4.6.2. Average functions

- Strain function

$$\varepsilon(t) = 9.821047696334.10^{-24}.t^9 - 5.87562064666.10^{-20}.t^8 + 1.466419386775959.10^{-16}.t^7 - 1.967948425335.10^{-13}.t^6 + 1.52291133354382.10^{-10}.t^5 - 6.7452158331146318.10^{-8}.t^4 + 1.57956315685248.10^{-5}.t^3 - 1.55050130427589.10^{-3}.t^2 - 2.3210007091027912.10^{-2}.t + 38.800 \quad (3.4.23)$$

- Stress function

$$\sigma(t) = 5.89562657252.10^{-23}.t^9 - 3.5021290564714.10^{-19}.t^8 + 8.837313814889449.10^{-16}.t^7 - 1.224964950874809.10^{-12}.t^6 + 1.006430542083158.10^{-9}.t^5 - 4.93032485793549.10^{-7}.t^4 + 1.3733133337954606.10^{-4}.t^3 - 1.928366035007075.10^{-2}.t^2 + 1.038885324332923.t + 285.08 \quad (3.4.24)$$

4.7. Young's modulus (E-modulus)

On table 3.4.7 is present a result is presented after processing, as the given values of the E-modulus are calculated as the tangent of the angle that the deformation curve makes with the abscissa axis of the σ-ε diagram and subsequently (the obtained values) are processed using the stochastic method and average method. Fig. 3.3.7 shows the relationship between the change in E-modulus and the time of action of the corrosion category.

Using the polynomial approximation [14-15,114-119] from fig. 3.3.7, functional dependence for the E-modulus in dependence of the time of the influence of the corrosion (in months) is established in functions or equations.

Table 3.4.7 Result after processing

	Stochastic results									
time, [months]	0	149	240	405	567	750	878	1050	1188	1256
E, [GPa]	211.50	168.10	153.50	148.70	156.20	151.90	131.60	153.50	184.50	232.00
	Average results									
time, [months]	0	149	240	405	567	750	878	1050	1188	1256
E, [GPa]	193.10	166.20	159.10	143.60	156.30	164.40	136.50	157.50	181.20	206.60

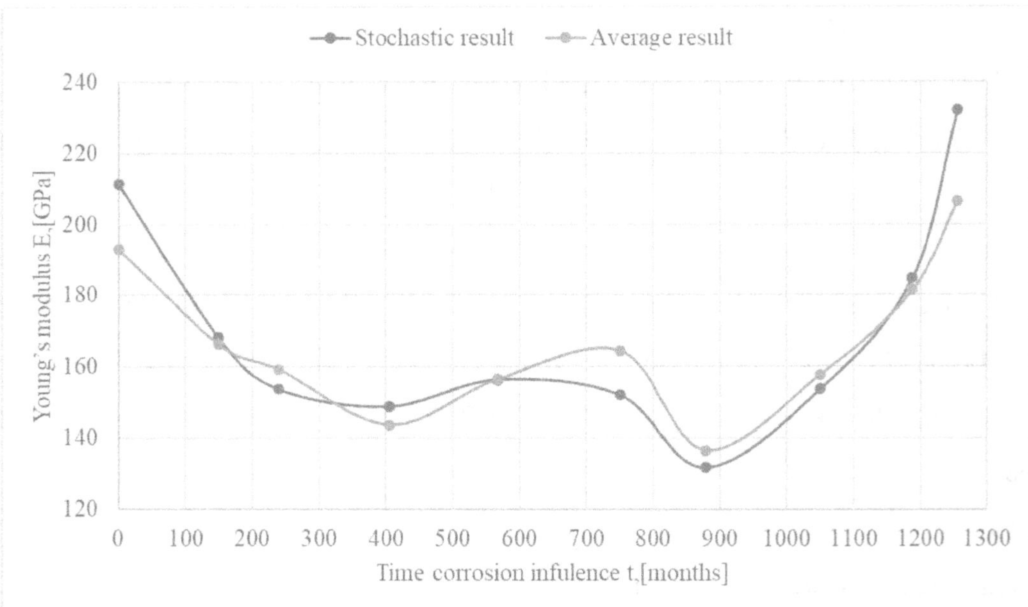

Figure 3.4.7 Graph of the variation of the Young's modulus (E-modulus) depending on the corrosion effect over time;

4.7.1. Stochastic functions

$$E(t) = 9.3322484212121.10^{-23}.t^{9} - 5.1408055294747.10^{-19}.t^{8} + 1.1798125659956.10^{-15}.t^{7}$$
$$- 1.4644361517818266.10^{-12}.t^{6} + 1.069174638359888.10^{-9}.t^{5}$$
$$- 4.67616975124223.10^{-7}.t^{4} + 21.184485325446.10^{-4}.t^{3} - 1.5147771145043547.10^{-2}.t^{2} \quad (3.4.25)$$
$$+ 0.45139469801009.t + 211.50$$

4.7.2. Average functions

$$E(t) = 1.210846433317.10^{-22}.t^{9} - 6.5883088526265.10^{-19}.t^{8} + 1.4795174417697.10^{-15}.t^{7}$$
$$- 11.7716763042046021.10^{-12}.t^{6} + 1.2215529575445629.10^{-9}.t^{5}$$
$$- 4.8955626487618.10^{-7}.t^{4} + 1.0966019623113669.10^{-4}.t^{3} - 1.2127838577855.10^{-2}.t^{2} \quad (3.4.26)$$
$$+ 0.32425555561921038.t + 193.10$$

5. Corrosion category C5

Corrosive category C5 is determined by the ISO 12944-5 standard, to be applied in industrial areas with high humidity and aggressive environments. Buildings and areas with permanent condensation and pollution over 50% of pollution in cities.

5.1. Point of the proportional limit

On table 3.5.1 is present a result after processing by stochastic method and average method. In fig. 3.5.1(a) shows the relationship between the change in relative strain versus the action time of the corrosion category, and in fig. 3.5.1(b) shows the relationship between the change in strength over time due to the corrosion effect of category.

Table 3.5.1 Result after processing [119]

	Stochastic results									
time, [months]	0	59	96	162	227	300	351	420	475	502
ε, [mm]	2.070	2.521	2.727	2.595	2.151	2.870	2.432	2.296	1.414	1.331
σ, [MPa]	218.91	211.90	209.30	193.00	224.00	218.00	213.40	234.90	173.90	205.90
	Average results									
time, [months]	0	59	96	162	227	300	351	420	475	502
ε, [mm]	2.130	2.543	2.619	2.664	2.039	2.511	2.423	1.907	1.299	1.282
σ, [MPa]	205.68	211.30	208.40	191.30	212.40	206.40	220.50	200.20	156.90	176.60

Using the polynomial approximation [14-15,114-119] from fig. 3.5.1, functional dependence for the strain and stress point in dependence of the time of the influence of the corrosion (in months) is established in functions or equations.

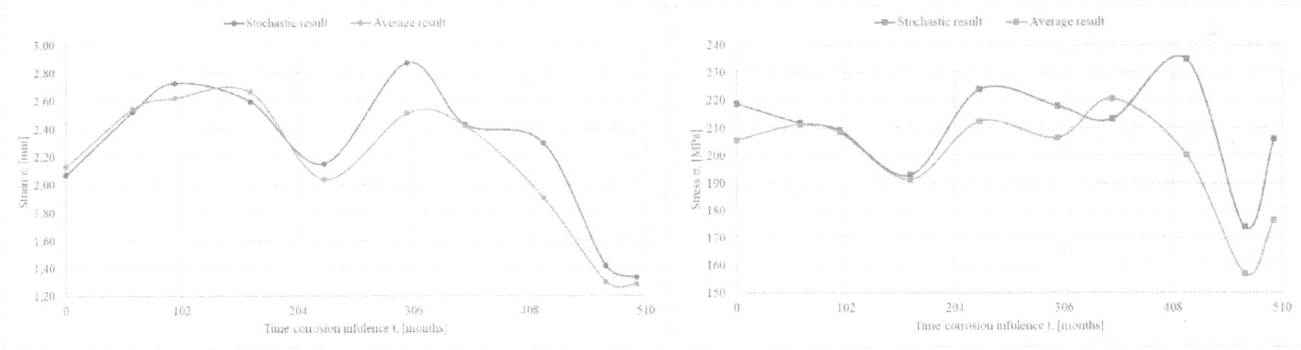

(a) (b)

Figure 3.5.1 (a) Graph of the variation of the strain deformation depending on the corrosion effect over time; [119]; (b) Graph of the change of stress strength depending on the corrosion effect over time [119];

5.1.1. Stochastic functions

- Strain function [119]

$$\begin{aligned}\varepsilon(t) &= 2.303825882465.10^{-20}.t^9 - 5.2805195971169.10^{-17}.t^8 + 5.05761608832997.10^{-14}.t^7 \\ &- 2.622599343527687.10^{-11}.t^6 + 7.9718377259304.10^{-9}.t^5 - 1.4368115220020867.10^{-6}.t^4 \\ &+ 1.47985401261478.10^{-4}.t^3 - 7.9072166587308354.10^{-3}.t^2 \\ &+ 1.7426975091279229.10^{-1}.t + 2.070\end{aligned} \quad (3.5.1)$$

- Stress function [119]

$$\begin{aligned}\sigma(t) &= - 5.5145431358569.10^{-20}.t^9 + 2.5608842459536787.10^{-16}.t^8 - 3.775781846.10^{-13}.t^7 \\ &+ 2.68939688 75467.10^{-10}.t^6 - 1.0543060686795274.10^{-7}.t^5 \\ &+ 2.3503097910172139.10^{-5}.t^4 - 2.8925352263123785.10^{-3}.t^3 \\ &+ 1.7779413999189486.10^{-1}.t^2 - 4.2662227549317757.t + 218.91\end{aligned} \quad (3.5.2)$$

5.1.2. Average functions

- Strain function [119]

$$\begin{aligned}\varepsilon(t) &= 1.405862222215695.10^{-20}.t^9 - 3.3329048048625.10^{-17}.t^8 + 3.31441231065.10^{-14}.t^7 \\ &- 1.791474826573546.10^{-11}.t^6 + 5.698303381844.10^{-9}.t^5 - 1.0786578046151475.10^{-6}.t^4 + \\ &1.1703719155662423.10^{-4}.t^3 - 6.600389797149427.10^{-3}.t^2 + 1.52991976542266.10^{-1}.t \\ &+ 2.130\end{aligned} \quad (3.5.3)$$

- Stress function [119]

$$\sigma(t) = -6.5171115293.10^{-19}.t^9 + 1.553746294399.10^{-15}.t^8 - 1.5503952313586989.10^{-12}.t^7 + 8.3993227295071 1.10^{-10}.t^6 - 2.67847511071652.10^{-7}.t^5 + 5.0880285508636739.10^{-5}.t^4 - 5.5310730395416704.10^{-3}.t^3 + 3.0766527852017855.10^{-1}.t^2 - 6.5463151111927704.t + 205.68 \quad (3.5.4)$$

5.2. Elastic limit point

On table 3.5.2 is present a result after processing by stochastic method and average method. In fig. 3.5.2(a) shows the relationship between the change in relative strain versus the action time of the corrosion category, and in fig. 3.5.2(b) shows the relationship between the change in strength over time due to the corrosion effect of category.

Table 3.5.2 Result after processing

	Stochastic results									
time, [months]	0	59	96	162	227	300	351	420	475	502
ε, [mm]	2.260	2.955	3.024	2.997	2.475	3.351	2.894	2.708	1.626	1.553
σ, [MPa]	292.80	293.90	276.50	283.50	281.70	282.40	286.80	283.90	233.90	232.90
	Average results									
time, [months]	0	59	96	162	227	300	351	420	475	502
ε, [mm]	2.691	2.967	3.055	3.108	2.378	2.930	2.827	2.225	1.516	1.495
σ, [MPa]	284.00	277.30	276.60	264.10	265.00	265.10	288.50	248.00	211.20	232.20

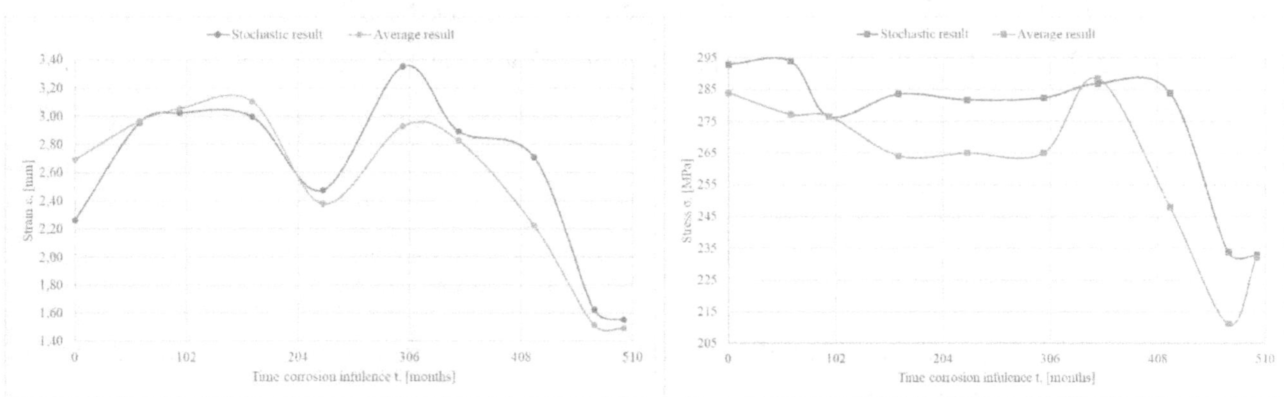

(a) (b)

Figure 3.5.2 (a) Graph of the variation of the strain deformation depending on the corrosion effect over time; (b) Graph of the change of stress strength depending on the corrosion effect over time;

Using the polynomial approximation [14-15,114-119] from fig. 3.5.2, functional dependence for the strain and stress point in dependence of the time of the influence of the corrosion (in months) is established in functions or equations.

5.2.1. Stochastic functions

- Strain function

$$\begin{aligned}\varepsilon(t) = {} & 2.8426519364712.10^{-20}.t^9 - 6.55926110039.10^{-17}.t^8 + 6.33636515237307.10^{-14}.t^7 \\ & - 3.3224454596756724.10^{-11}\ t^6 + 1.02492184570444.10^{-8}.t^5 \\ & - 1.8845603539237625.10^{-6}.t^4 + 1.9950686157663112.10^{-4}.t^3 \\ & - 1.10603691948267.10^{-2}.t^2 + 0.25395238226133615.t + 2.260\end{aligned} \qquad (3.5.5)$$

- Stress function

$$\begin{aligned}\sigma(t) = {} & 2.5327645587.10^{-19}.t^9 - 5.89970164214.10^{-16}.t^8 + 5.8550172372216495.10^{-13}.t^7 \\ & - 3.224397822154474.10^{-10}.t^6 + 1.072615588965586.10^{-7}.t^5 - 2.1907634599984228.10^{-5}.t^4 \\ & + 2.6529100599666775.10^{-3}.t^3 - 0.17092489964542573.t^2 + 4.2753220249076485.t \\ & + 292.84\end{aligned} \qquad (3.5.6)$$

5.2.2. Average functions

- Strain function

$$\begin{aligned}\varepsilon(t) = {} & 1.6346165382995.10^{-20}.t^9 - 3.873218686918.10^{-17}.t^8 + 3.8489667615838.10^{-14}.t^7 \\ & - 2.078244050538627.10^{-11}.t^6 + 6.5999112109914843.10^{-9}.t^5 - 1.24602757723875.10^{-6}.t^4 \\ & + 1.3453888548702509.10^{-4}.t^3 - 7.5071776965180718.10^{-3}.t^2 + 0.1685339209147275.t \\ & + 2.691\end{aligned} \qquad (3.5.7)$$

- Stress function

$$\begin{aligned}\sigma(t) = {} & - 6.376837746876.10^{-19}.t^9 + 1.466134711636.10^{-15}.t^8 - 1.4082708080029.10^{-12}.t^7 \\ & + 7.3377499723288272.10^{-10}.t^6 - 2.25320934961452.10^{-7}.t^5 + 4.141970482459498.10^{-5}.t^4 \\ & - 4.4053597440227854.10^{-3}.t^3 + 2.4465374254552025.10^{-1}\ t^2 - 5.458248298558189\ t \\ & + 284.00\end{aligned} \qquad (3.5.8)$$

5.3. Yield point

On table 3.5.3 is present a result after processing by stochastic method and average method. In fig. 3.5.3(a) shows the relationship between the change in relative strain versus the action time of the corrosion category, and in fig. 3.5.3(b) shows the relationship between the change in strength over time due to the corrosion effect of category.

Table 3.5.3 Result after processing [115, 117]

	Stochastic results									
time, [months]	0	59	96	162	227	300	351	420	475	502
ε, [mm]	5.730	5.154	6.83	6.172	3.625	4.678	4.376	3.594	2.254	1.998
σ, [MPa]	358.67	332.90	329.90	339.80	354.10	337.40	352.80	331.90	327.40	312.90
	Average results									
време, [months]	0	59	96	162	227	300	351	420	475	502
ε, [mm]	6.010	5.547	6.664	6.363	3.863	4.459	4.285	3.358	2.065	1.921
σ, [MPa]	344.79	326.20	347.10	352.10	329.80	337.20	361.10	305.20	280.00	317.10

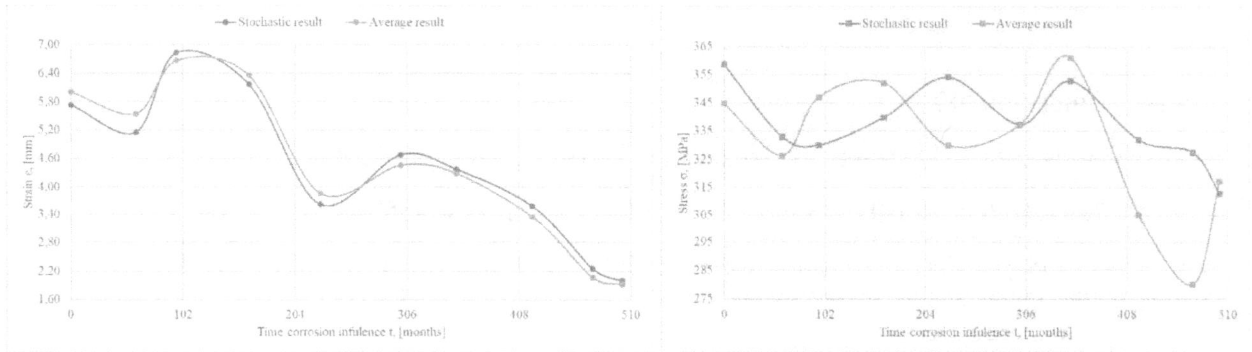

(a) (b)

Figure 3.5.3 (a) Graph of the variation of the strain deformation depending on the corrosion effect over time; [115]; (b) Graph of the change of stress strength depending on the corrosion effect over time [117];

Using the polynomial approximation [14-15,114-119] from fig. 3.5.3, functional dependence for the strain and stress point in dependence of the time of the influence of the corrosion (in months) is established in functions or equations.

5.3.1. Stochastic functions

- Strain function [115]

$$\varepsilon(t) = 3.5395616558633.10^{-20}.t^{9} - 8.1855457212498.10^{-17}.t^{8} + 7.8808214211733.10^{-14}.t^{7} \\ - 4.077254770489.10^{-11}.t^{6} + 1.218634625446397.10^{-8}.t^{5} - 2.0986031708782749.10^{-6}.t^{4} \qquad (3.5.9)$$

$+ 1.9482229298127139.10^{-4}.t^3 - 8.300722248339356.10^{-3}.t^2 + 1.11169025465154.10^{-1}.t + 5.730$

- Stress function [117]

$$\sigma(t) = - 7.56892169617.10^{-19}.t^9 + 1.72363431053.10^{-15}.t^8 - 1.639278511756799.10^{-12}.t^7 + 8.440111640511105.10^{-10}.t^6 - 2.5497666157232.10^{-7}.t^5 + 4.5806869996493698.10^{-5}.t^4 - 4.7357436324078021.10^{-3}.t^3 + 2.58371304396951.10^{-1}.t^2 - 6.052108832116124.t + 358.67 \quad (3.5.10)$$

5.3.2. Average functions

- Strain function [115]

$$\varepsilon(t) = 3.0881804265966.10^{-20}.t^9 - 7.2543216386226.10^{-17}.t^8 + 7.12055369502.10^{-14}.t^7 - 3.77505473702118.10^{-11}.t^6 + 1.16514609920839.10^{-8}.t^5 - 2.0987488158225203.10^{-6}.t^4 + 2.088381992553756.10^{-4}.t^3 - 1.006799064325816.10^{-3}.t^2 + 1.732127655383256.10^{-1}.t + 6.010 \quad (3.5.11)$$

- Stress function [117]

$$\sigma(t) = - 3.80344342082.10^{-19}.t^9 + 8.33126208555595.10^{-16}.t^8 - 7.5828820601766.10^{-13}.t^7 + 3.73992477095.10^{-10}.t^6 - 1.097735651725519.10^{-7}.t^5 + 1.9906955771755023.10^{-5}.t^4 - 2.230133655368107.10^{-3}.t^3 + 1.43786615063275.10^{-1}.t^2 - 4.0311064232796738.t + 344.79 \quad (3.5.12)$$

5.4. Lower yield point

On table 3.5.4 is present a result after processing by stochastic method and average method. In fig. 3.5.4(a) shows the relationship between the change in relative strain versus the action time of the corrosion category, and in fig. 3.5.4(b) shows the relationship between the change in strength over time due to the corrosion effect of category.

Using the polynomial approximation [14-15,114-119] from fig. 3.5.4, functional dependence for the strain and stress point in dependence of the time of the influence of the corrosion (in months) is established in functions or equations.

Table 3.5.4 Result after processing [118]

	Stochastic results									
time, [months]	0	59	96	162	227	300	351	420	475	502
ε, [mm]	8.760	9.034	9.939	9.120	5.490	6.125	5.554	5.295	2.654	2.274
σ, [MPa]	363.43	347.40	366.40	405.30	357.90	372.90	369.80	345.50	377.90	363.90
	Average results									
time, [months]	0	59	96	162	227	300	351	420	475	502
ε, [mm]	8.950	8.366	9.836	9.174	5.008	5.569	5.339	4.173	2.399	2.132
σ, [MPa]	359.19	343.70	365.90	376.90	352.90	358.60	379.90	316.10	295.70	339.40

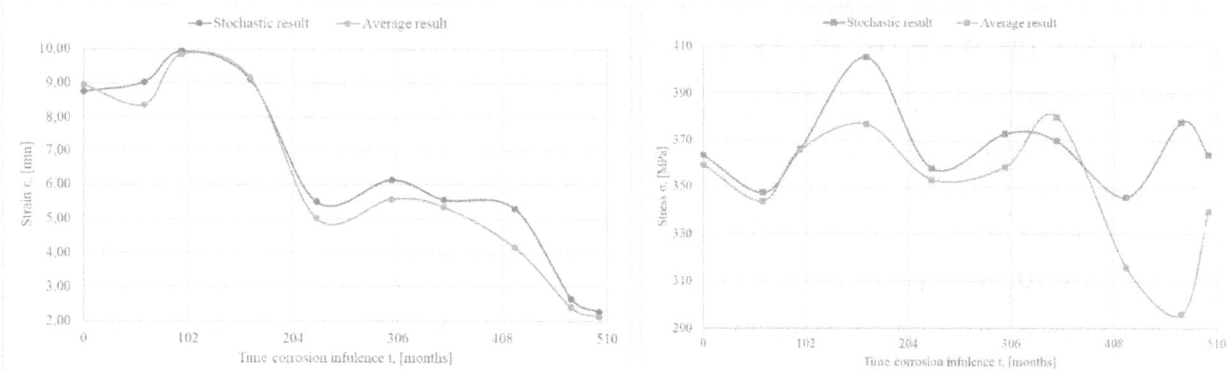

(a) (b)

Figure 3.5.4 (a) Graph of the variation of the strain deformation depending on the corrosion effect over time; [118]; (b) Graph of the change of stress strength depending on the corrosion effect over time [118];

5.4.1. Stochastic functions

- Strain function [118]

$$\begin{aligned}\varepsilon (t) = {} & 5.901376749657.10^{-20}.t^{9} - 1.368455872003.10^{-16}.t^{8} + 1.3281593738738.10^{-14}.t^{7} \\ & - 6.9861510161558375.10^{-11}.t^{6} + 2.152912467217.10^{-8}.t^{5} - 3.9176048178681465.10^{-6}.t^{4} \\ & + 4.0281817310731596.10^{-4}.t^{3} - 2.102463502790603.10^{-3}.t^{2} + 4.31279059704405.10^{-1}.t \\ & + 8.760\end{aligned} \quad (3.5.13)$$

- Stress function [118]

$$\sigma (t) = 6.954828694154.10^{-19}.t^9 - 1.7326841437104.10^{-15}.t^8 + 1.805135679667.10^{-12}.t^7 - 1.0185857581131115.10^{-9}.t^6 + 3.3666868039223373.10^{-7}.t^5 - 6.5701788587359.10^{-5}.t^4 + 7.225545178276475.10^{-3}.t^3 - 3.9608021238996538.10^{-1}.t^2 + 8.0154059884162265.t + 363.43 \quad (3.5.14)$$

5.4.2. Average functions

- Strain function [118]

$$\varepsilon (t) = 4.681784265325.10^{-20}.t^9 - 1.102894288038.10^{-16}.t^8 + 1.08637198817588.10^{-13}.t^7 - 5.78560939653.10^{-11}.t^6 + 1.7964579789200959.10^{-8}.t^5 - 3.263342510049432.10^{-6}.t^4 + 3.28999227316859.10^{-4}.t^3 - 1.624946359776218.10^{-2}.t^2 + 2.9315929548303077.10^{-1}.t + 8.950 \quad (3.5.15)$$

- Stress function [118]

$$\sigma (t) = -3.6271506728959.10^{-19}.t^9 + 7.796098494554.10^{-15}.t^8 - 6.926139485682.10^{-13}.t^7 + 3.3131787527613485.10^{-10}.t^6 - 9.375894107011401.10^{-8}.t^5 + 1.63826126163827.10^{-5} .t^4 - 1.79300884021562.10^{-3}.t^3 + 1.1641100737387665.10^{-1}.t^2 - 3.3273950132025627.t + 359.19 \quad (3.5.16)$$

5.5. Ultimate strength point

On table 3.5.5 is present a result after processing by stochastic method and average method. In fig. 3.5.5(a) shows the relationship between the change in relative strain versus the action time of the corrosion category, and in fig. 3.5.5(b) shows the relationship between the change in strength over time due to the corrosion effect of category.

Using the polynomial approximation [14-15,114-119] from fig. 3.5.5, functional dependence for the strain and stress point in dependence of the time of the influence of the corrosion (in months) is established in functions or equations.

5.5.1. Stochastic functions

- Strain function [114]

$$\varepsilon (t) = 3.0511197949414.10^{-20}.t^9 - 7.6107763657397.10^{-17}.t^8 + 7.995456730331.10^{-14}.t^7 - 4.5972954240413.10^{-11}.t^6 + 1.5736520740782.10^{-8}.t^5 - 3.2661013387948908.10^{-6}.t^4 + 4.000747175415075.10^{-4}.t^3 - 2.6485976095034.10^{-2}.t^2 + 6.76913288856759.10^{-1}.t + 28.540 \quad (3.5.17)$$

- Stress function [114]

$$\sigma(t) = 2.1632760408248.10^{-19}.t^9 - 6.00189652095427.10^{-16}.t^8 + 6.8810106119932.10^{-13}.t^7 - 4.231212325599.10^{-10}.t^6 + 1.5119601779336627.10^{-7}.t^5 - 3.1706570323848069.10^{-5}.t^4 + 3.732934965321672.10^{-3}.t^3 - 2.18883163632775.10^{-1}.t^2 + 14.670895667683998.t + 460.91 \quad (3.5.18)$$

5.5.2. Average functions

- Strain function [114]

$$\varepsilon(t) = 4.452126698723358.10^{-20}.t^9 - 1.051478382688.10^{-16}.t^8 + 1.03681401951.10^{-13}.t^7 - 5.5079832555745539.10^{-11}.t^6 + 1.6938262555261338.10^{-8}.t^5 - 3.00475576609.10^{-6}.t^4 + 2.87190045307279.10^{-4}.t^3 - 1.23814007169599.10^{-2}.t^2 + 7.300419381613759.10^{-2}.t + 29.600 \quad (3.5.19)$$

- Stress function [114]

$$\sigma(t) = 8.1337849352421449.10^{-21}.t^9 - 34.789147895596.10^{-17}.t^8 + 8.26308648559.10^{-14}.t^7 - 6.5633524129034737.10^{-11}.t^6 + 2.7480019225421485.10^{-8}.t^5 - 6.2415681347014.10^{-6}.t^4 + 7.35257276757917.10^{-4}.t^3 - 3.8845250592370567.10^{-2}.t^2 + 0.553708461990625.t + 447.20 \quad (3.5.20)$$

Table 3.5.5 Result after processing [114]

	Stochastic results									
time, [months]	0	59	96	162	227	300	351	420	475	502
ε, [mm]	28.540	28.370	23.790	19.730	15.270	13.120	11.250	8.090	6.070	3.430
σ, [MPa]	460.910	448.90	450.90	474.80	444.00	446.90	449.90	413.90	433.90	451.90
	Average results									
time, [months]	0	59	96	162	227	300	351	420	475	502
ε, [mm]	29.600	23.410	23.440	19.440	12.470	12.250	12.000	8.890	4.550	3.550
σ, [MPa]	447.20	437.10	441.30	444.20	423.70	431.40	443.00	390.60	354.40	421.00

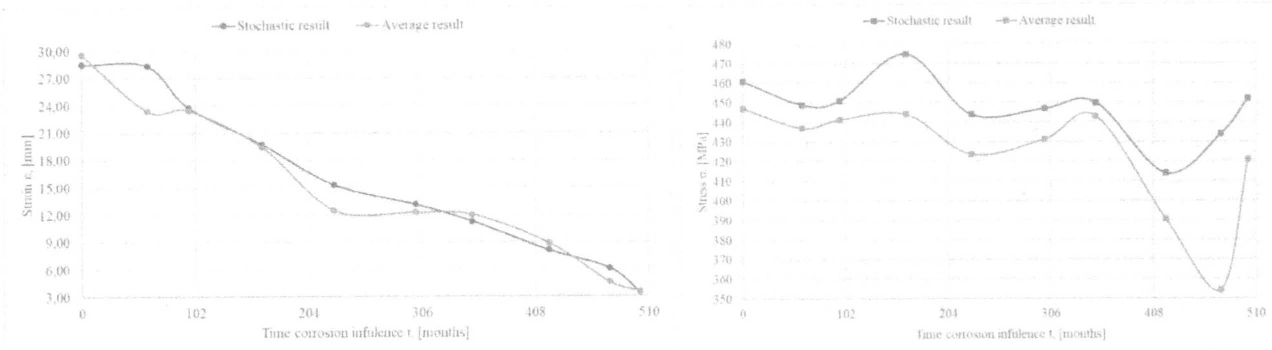

(a) (b)

Figure 3.5.5 (a) Graph of the variation of the strain deformation depending on the corrosion effect over time; [114]; (b) Graph of the change of stress strength depending on the corrosion effect over time [114];

5.6.Fracture point

On table 3.5.6 is present a result after processing by stochastic method and average method. In fig. 3.5.6(a) shows the relationship between the change in relative strain versus the action time of the corrosion category, and in fig. 3.5.6(b) shows the relationship between the change in strength over time due to the corrosion effect of category.

Table 3.5.6 Result after processing

	Stochastic results									
time, [months]	0	59	96	162	227	300	351	420	475	502
ε, [mm]	40.410	36.080	28.360	23.420	17.600	16.640	12.990	10.230	6.750	5.170
σ, [MPa]	286.41	282.70	284.40	297.40	288.90	285.90	305.90	289.90	287.90	322.90
	Average results									
time, [months]	0	59	96	162	227	300	351	420	475	502
ε, [mm]	38.800	29.180	28.250	23.020	15.120	14.370	14.160	10.280	5.630	4.310
σ, [MPa]	285.08	284.90	290.50	304.90	292.70	298.30	300.30	263.70	231.00	295.30

Using the polynomial approximation [14-15,114-119] from fig. 3.5.6, functional dependence for the strain and stress point in dependence of the time of the influence of the corrosion (in months) is established in functions or equations.

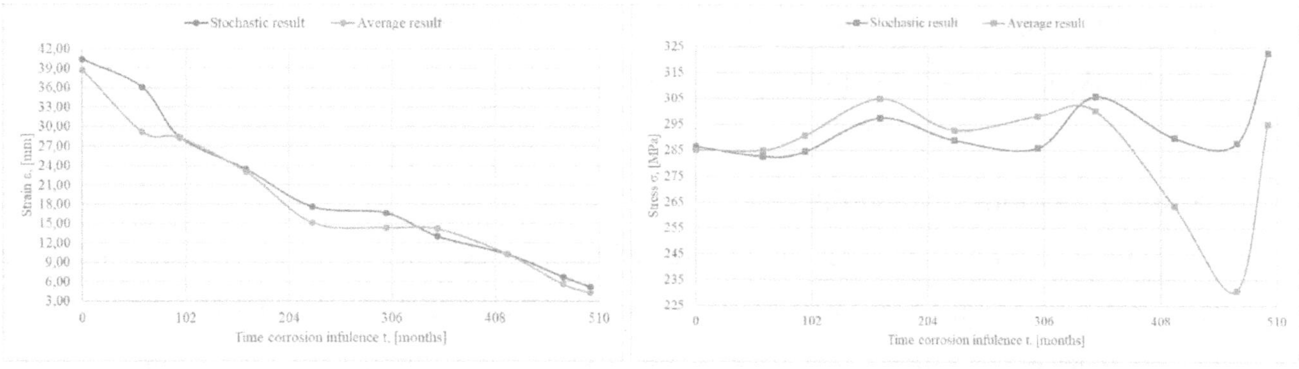

(a) (b)

Figure 3.5.6 (a) Graph of the variation of the strain deformation depending on the corrosion effect over time; (b) Graph of the change of stress strength depending on the corrosion effect over time;

5.6.1. Stochastic functions

- Strain function

$$\begin{aligned}\varepsilon(t) = {} & 1.121722733505.10^{-19}.t^9 - 2.6542236307297.10^{-16}.t^8 + 2.64238257203.10^{-13}.t^7 \\ & - 1.4378330910674545.10^{-10}.t^6 + 4.6485814336253.10^{-8}.t^5 \\ & - 9.0823979718546173.10^{-6}.t^4 + 1.0399255199644896.10^{-3}.t^3 \\ & - 6.32026995986336.10^{-2}.t^2 + 1.429926810584945.t + 40.410\end{aligned} \quad (3.5.21)$$

- Stress function

$$\begin{aligned}\sigma(t) = {} & -2.50181058097.10^{-19}.t^9 + 5.39924701797.10^{-16}.t^8 - 4.766453729329362.10^{-13}.t^7 \\ & + 2.2171478703994058.10^{-10}.t^6 - 5.841096031020797.10^{-8}.t^5 \\ & + 8.752769401051822.10^{-6}.t^4 - 7.2052000540210269.10^{-4}.t^3 \\ & + 3.1011640010836602.10^{-2}.t^2 - 0.61399360949256.t + 386.41\end{aligned} \quad (3.5.22)$$

5.6.2. Average functions

- Strain function

$$\begin{aligned}\varepsilon(t) = {} & 3.71315416245.10^{-20}.t^9 - 8.877812799966.10^{-17}.t^8 + 8.8535428931413.10^{-14}.t^7 \\ & - 4.74634043885.10^{-11}.t^6 + 1.4664339303695967.10^{-8}.t^5 - 2.5900015503000213.10^{-6}.t^4 \\ & + 2.41120800413921.10^{-4}.t^3 - 9.304359292353833.10^{-3}.t^2 - 6.878347415695457.10^{-2}.t \\ & + 38.800\end{aligned} \quad (3.5.23)$$

- Stress function

$$\begin{aligned}\sigma(t) = {} & 2.272202318029.10^{-19}.t^9 - 5.3881456863749.10^{-16}.t^8 \\ & + 5.427946236246857.10^{-13}.t^7 - 3.0039025083643011.10^{-10}.t^6\end{aligned} \quad (3.5.24)$$

$+ 9.854559049698.10^{-8}.t^5 - 1.927726822195026.10^{-5}.t^4 + 2.1439869657273397.10^{-3}.t^3$

$- 1.2015716030364.10^{-1}.t^2 + 2.5811990087078396.t + 385.08$

5.7. Young's modulus (E-modulus)

On table 3.5.7 is present a result is presented after processing, as the given values of the E-modulus are calculated as the tangent of the angle that the deformation curve makes with the abscissa axis of the σ-ε diagram and subsequently (the obtained values) are processed using the stochastic method and average method. Fig. 3.3.7 shows the relationship between the change in E-modulus and the time of action of the corrosion category.

Using the polynomial approximation [14-15,114-119] from fig. 3.3.7, functional dependence for the E-modulus in dependence of the time of the influence of the corrosion (in months) is established in functions or equations.

5.7.1. Stochastic functions

$$E(t) = 3.576992534.10^{-19}.t^9 - 7.87270113279.10^{-16}.t^8 + 7.2168823729752788.10^{-13}.t^7 - 3.5765203719342446.10^{-10}.t^6 + 1.04180386257137.10^{-7}.t^5 - 1.8158852558327529.10^{-5}.t^4 + 1.8298121721211.10^{-3}.t^3 - 9.2746321474062879.10^{-2}.t^2 + 1.0610722942708066.t + 211.50 \quad (3.5.25)$$

5.7.2. Average functions

$$E(t) = 4.67236677777.10^{-19}.t^9 - 1.0168456149675.10^{-15}.t^8 + 9.134148167074259.10^{-13}.t^7 - 4.3754929685249855.10^{-10}.t^6 + 1.2068448361860669.10^{-7}.t^5 - 1.9344688860504.10^{-5}.t^4 + 1.7320037613201833.10^{-3}.t^3 - 7.6439495474527.10^{-2}.t^2 + 0.81224888942632756.t + 193.10 \quad (3.5.26)$$

Table 3.5.7 Result after processing

	Stochastic results									
time, [months]	0	59	96	162	227	300	351	420	475	502
E, [GPa]	211.50	168.10	153.50	148.70	156.20	151.90	131.60	153.50	184.50	232.00
	Average results									
time, [months]	0	59	96	162	227	300	351	420	475	502
E, [GPa]	193.10	166.20	159.10	143.60	156.30	164.40	136.50	157.50	181.20	206.60

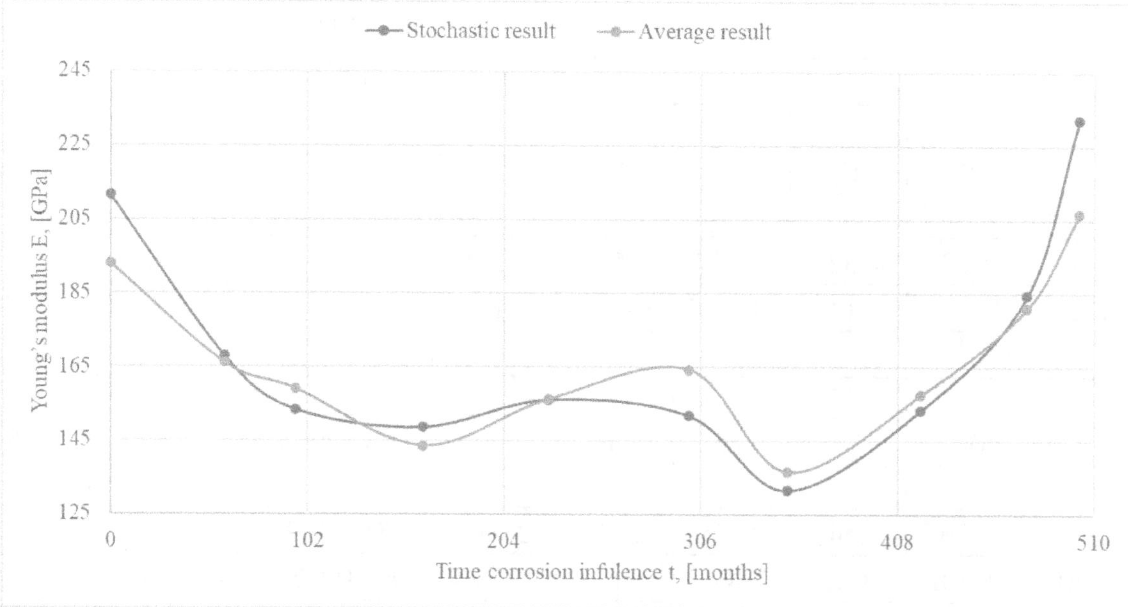

Figure 3.5.7 Graph of the variation of the Young's modulus (E-modulus) depending on the corrosion effect over time;

6. Corrosion category CX

The ISO 12944-5 standard has defined this category as a coastal zone with offshore installations and a high salt content. Buildings and structural facilities with a high degree of condensation and pollution over 75% of pollution in the chemical production of aggressive components.

6.1. Point of the proportional limit

On table 3.6.1 is present a result after processing by stochastic method and average method. In fig. 3.6.1(a) shows the relationship between the change in relative strain versus the action time of the corrosion category, and in fig. 3.6.1(b) shows the relationship between the change in strength over time due to the corrosion effect of category.

Using the polynomial approximation [14-15,114-119] from fig. 3.6.1, functional dependence for the strain and stress point in dependence of the time of the influence of the corrosion (in months) is established in functions or equations.

Table 3.6.1 Result after processing [119]

	Stochastic results									
time, [months]	0	17	27	46	65	86	100	120	136	143
ε, [mm]	2.070	2.521	2.727	2.595	2.151	2.87	2.432	2.296	1.414	1.331
σ, [MPa]	218.91	211.90	209.30	193.00	224.00	218.00	213.40	234.90	173.90	205.90
	Average results									
time, [months]	0	17	27	46	65	86	100	120	136	143
ε, [mm]	2.130	2.543	2.619	2.664	2.039	2.511	2.423	1.907	1.299	1.282
σ, [MPa]	205.68	211.30	208.40	191.30	212.40	206.40	220.50	200.20	156.90	176.60

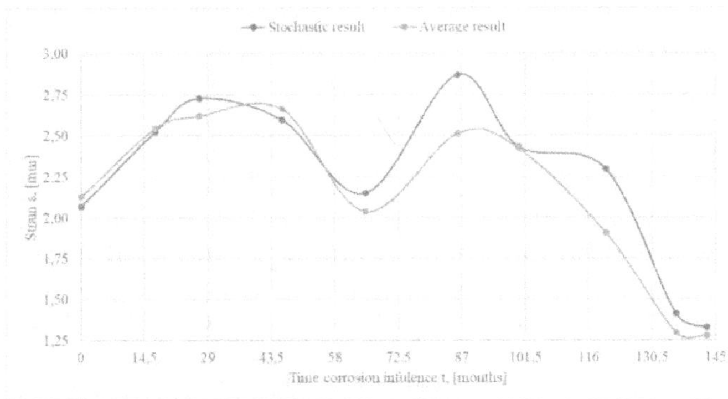

(a)

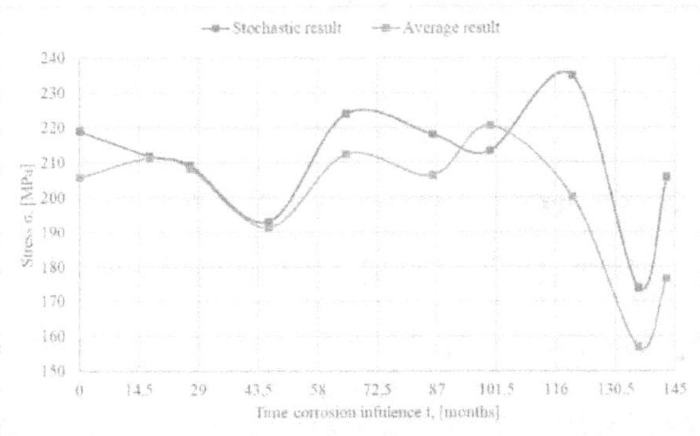

(b)

Figure 3.6.1 (a) Graph of the variation of the strain deformation depending on the corrosion effect over time; [119]; (b) Graph of the change of stress strength depending on the corrosion effect over time [119];

6.1.1. Stochastic functions

- Strain function [119]

$$\begin{aligned}\varepsilon(t) = {} & 1.8086755307.10^{-15}.t^{9} - 1.182307794856.10^{-12}.t^{8} + 3.2284281246556614.10^{-10}.t^{7} \\ & - 4.7705102281042225.10^{-8}.t^{6} + 4.129592346897562.10^{-6}.t^{5} \\ & - 2.1179274597436754.10^{-4}.t^{4} + 6.201271277208.10^{-3}.t^{3} \\ & - 9.4141561719018427.10^{-2}.t^{2} + 5.9081028920008327.10^{-1}.t + 2.070\end{aligned} \qquad (3.6.1)$$

- Stress function [119]

$$
\begin{aligned}
\sigma(t) = &- 2.056173235656.10^{-15}.t^{9} + 4.246950743578.10^{-12}.t^{8} - 2.0029872916941.10^{-9}.t^{7} \\
&+ 4.287763714484875.10^{-7}.t^{6} - 4.9364318210086465.10^{-5}.t^{5} \\
&+ 3.194821422438193.10^{-3}.t^{4} - 1.1335523090626587.10^{-1}.t^{3} + 1.9998299464220755.t^{2} \\
&- 13.785160790774738.t + 218.91
\end{aligned} \tag{3.6.2}
$$

6.1.2. Average functions

- Strain function [119]

$$
\begin{aligned}
\varepsilon(t) = &\ 1.080330914186.10^{-15}.t^{9} - 7.3069868356.10^{-13}.t^{8} + 2.0727227316024278.10^{-10}.t^{7} \\
&- 3.194837601428119.10^{-8}.t^{6} + 2.8968663309163941.10^{-6}.t^{5} \\
&- 1.5624622753412657.10^{-4}.t^{4} + 4.828306694534999.10^{-3}.t^{3} \\
&- 7.7563219469705.10^{-2}.t^{2} + 5.1382636018591388.10^{-1}.t + 2.130
\end{aligned} \tag{3.6.3}
$$

- Stress function [119]

$$
\begin{aligned}
\sigma(t) = &- 5.0686081188735.10^{-14}.t^{9} + 3.449183129829458.10^{-11}.t^{8} - 9.8203041081.10^{-9}.t^{7} \\
&+ 1.5173825643904976.10^{-6}.t^{6} - 1.3794435987529.10^{-4}.t^{5} + 7.466244077719489.10^{-3}.t^{4} \\
&- 2.3112681692345116.10^{-1}.t^{3} + 3.6593940097517721.t^{2} - 22.175009798455623.t \\
&+ 205.68
\end{aligned} \tag{3.6.4}
$$

6.2. Elastic limit point

On table 3.6.2 is present a result after processing by stochastic method and average method. In fig. 3.6.2(a) shows the relationship between the change in relative strain versus the action time of the corrosion category, and in fig. 3.6.2(b) shows the relationship between the change in strength over time due to the corrosion effect of category.

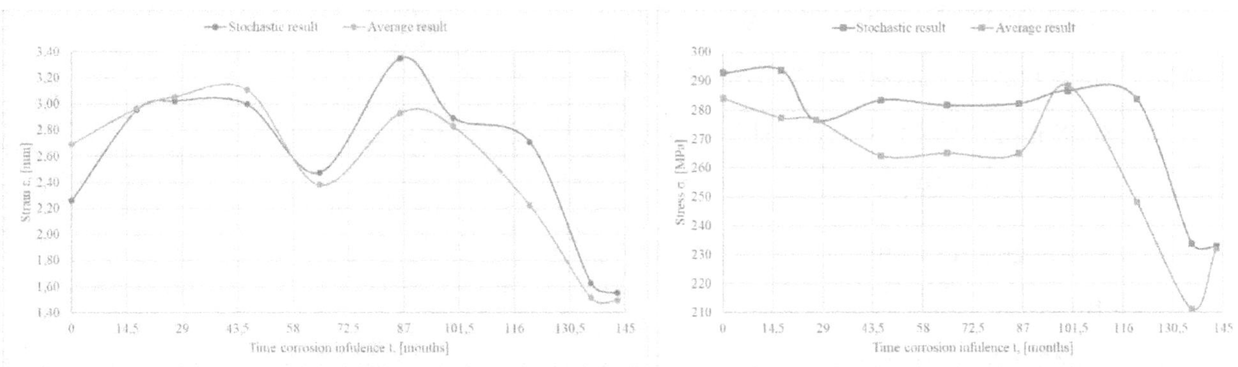

(a) (b)

Figure 3.6.2 (a) Graph of the variation of the strain deformation depending on the corrosion effect over time; (b) Graph of the change of stress strength depending on the corrosion effect over time;

Table 3.6.2 Result after processing

	Stochastic results									
time, [months]	0	17	27	46	65	86	100	120	136	143
ε, [mm]	2.260	2.955	3.024	2.997	2.475	3.351	2.894	2.708	1.626	1.553
σ, [MPa]	292.80	293.90	276.50	283.50	281.70	282.40	286.80	283.90	233.90	232.90
	Average results									
time, [months]	0	17	27	46	65	86	100	120	136	143
ε, [mm]	2.691	2.967	3.055	3.108	2.378	2.930	2.827	2.225	1.516	1.495
σ, [MPa]	284.00	277.30	276.60	264.10	265.00	265.10	288.50	248.00	211.20	232.20

.Using the polynomial approximation [14-15,114-119] from fig. 3.6.2, functional dependence for the strain and stress point in dependence of the time of the influence of the corrosion (in months) is established in functions or equations.

6.2.1. Stochastic functions

- Strain function

$$\begin{aligned}\varepsilon(t) = {} & 2.232805065286.10^{-15}.t^9 - 1.4697902684483.10^{-12}.t^8 + 4.04959662629585.10^{-10}.t^7 \\ & - 6.0543768582233846.10^{-8}.t^6 + 5.3232999600423.10^{-6}.t^5 - 2.7886702858490481.10^{-4}.t^4 \\ & + 8.4080260301543207.10^{-3}.t^3 - 0.13278420574857744.t^2 + 0.87053361107796856.t \\ & + 2.260\end{aligned} \tag{3.6.5}$$

- Stress function

$$\begin{aligned}\sigma(t) = {} & 1.99294793353157.10^{-14}.t^9 - 1.3293152202095.10^{-11}.t^8 + 3.779514509702.10^{-9}.t^7 \\ & - 5.965673870660925.10^{-7}.t^6 + 5.6900068864328432.10^{-5}.t^5 \\ & - 3.3329200868855626.10^{-3}.t^4 + 1.157616989079.10^{-1}.t^3 - 2.1394812267283814.t^2 \\ & + 15.3641355382647.t + 292.84\end{aligned} \tag{3.6.6}$$

6.2.2. Average functions

- Strain function

$$\begin{aligned}\varepsilon(t) = {} & 1.2558844929566.10^{-15}.t^9 - 8.48981687654656.10^{-13}.t^8 + 2.406450041.10^{-10}.t^7 \\ & - 3.705213498259799.10^{-8}.t^6 + 3.3540438851145869.10^{-6}.t^5 \\ & - 1.8040491011064473.10^{-4}.t^4 + 5.546424637911574.10^{-3}.t^3\end{aligned} \tag{3.6.7}$$

$- 8.8112613114054467.10^{-2}.t^2 + 0.5645669245612107.t + 2.691$

- Stress function

$$\sigma(t) = - 5.106565065.10^{-14}.t^9 + 3.3524736051328.10^{-11}.t^8 - 9.192213189485567.10^{-9}.t^7$$
$$+ 1.3667012322750557.10^{-6}.t^6 - 1.196958723828088.10^{-4}.t^5$$
$$+ 6.2720452370986345.10^{-3}.t^4 - 1.9005726592976033.10^{-1}.t^3 \quad (3.6.8)$$
$$+ 3.0067514578006573.t^2 - 19.131748259797199.t + 284.00$$

6.3. Yield point

On table 3.6.3 is present a result after processing by stochastic method and average method. In fig. 3.6.3(a) shows the relationship between the change in relative strain versus the action time of the corrosion category, and in fig. 3.6.3(b) shows the relationship between the change in strength over time due to the corrosion effect of category.

Using the polynomial approximation [14-15,114-119] from fig. 3.6.3, functional dependence for the strain and stress point in dependence of the time of the influence of the corrosion (in months) is established in functions or equations.

Table 3.6.3 Result after processing [115, 117]

	Stochastic results									
time, [months]	0	17	27	46	65	86	100	120	136	143
ε, [mm]	5.730	5.154	6.83	6.172	3.625	4.678	4.376	3.594	2.254	1.998
σ, [MPa]	358.67	332.90	329.90	339.80	354.10	337.40	352.80	331.90	327.40	312.90
	Average results									
time, [months]	0	17	27	46	65	86	100	120	136	143
ε, [mm]	6.010	5.547	6.664	6.363	3.863	4.459	4.285	3.358	2.065	1.921
σ, [MPa]	344.79	326.20	347.10	352.10	329.80	337.20	361.10	305.20	280.00	317.10

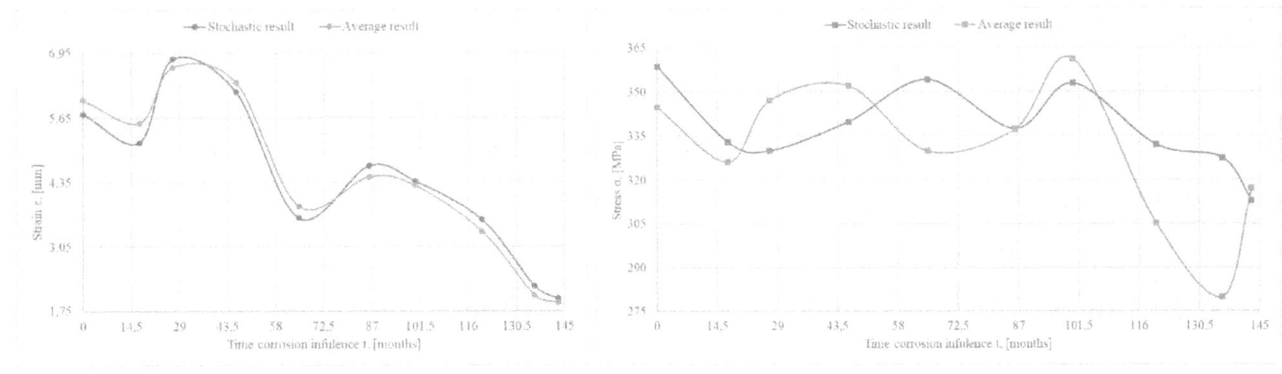

(a) (b)

Figure 3.6.3 (a) Graph of the variation of the strain deformation depending on the corrosion effect over time; [115]; (b) Graph of the change of stress strength depending on the corrosion effect over time [117];

6.3.1. Stochastic functions

- Strain function [115]

$$\varepsilon(t) = 2.63880457927.10^{-15}.t^{9} - 1.7364520418729845.10^{-12}.t^{8} + 4.751148632458.10^{-10}.t^{7} - 6.971707785432855.10^{-8}.t^{6} + 5.889885472045699.10^{-6}.t^{5} - 2.848621791766.10^{-4}.t^{4} + 7.3245849354254874.10^{-3}.t^{3} - 8.334703539523712.10^{-2}.t^{2} + 2.6202141008001.10^{-1}.t + 5.730 \quad (3.6.9)$$

- Stress function [117]

$$\sigma(t) = -6.00807470079.10^{-14}.t^{9} + 3.9013623826.10^{-11}.t^{8} - 1.057782832424982.10^{-8}.t^{7} + 1.5521379886814497.10^{-6}.t^{6} - 1.335810471393186.10^{-4}.t^{5} + 6.8332400769025529.10^{-3}.t^{4} - 2.0107006055750659.10^{-1}.t^{3} + 3.1227174288776047.t^{2} - 20.871710673121758.t + 358.67 \quad (3.6.10)$$

6.3.2. Average functions

- Strain function [115]

$$\varepsilon(t) = 2.30061415140044.10^{-15}.t^{9} - 1.5395694680889.10^{-12}.t^{8} + 4.301663509.10^{-10}.t^{7} - 6.483824399852877.10^{-8}.t^{6} + 5.6779740307026982.10^{-6}.t^{5} - 2.89158316241407.10^{-4}.t^{4} + 8.08037163531311.10^{-3}.t^{3} - 1.079444507062307.10^{-1}.t^{2} + 5.0128769248895777.10^{-1}.t + 6.010 \quad (3.6.11)$$

- Stress function [117]

$$\sigma(t) = -3.22282348483.10^{-14}.t^9 + 2.0331408939165.10^{-11}.t^8 - 5.337864366861646.10^{-9}.t^7 + 7.6057090671180656.10^{-7}.t^6 - 6.45394553576.10^{-5}.t^5 + 3.3775334322839546.10^{-3}.t^4 - 1.084994365699714.10^{-1}.t^3 + 1.9851331705231345.t^2 - 15.647011259819337.t + 344.79 \quad (3.6.12)$$

6.4. Lower yield point

On table 3.6.4 is present a result after processing by stochastic method and average method. In fig. 3.6.4(a) shows the relationship between the change in relative strain versus the action time of the corrosion category, and in fig. 3.6.4(b) shows the relationship between the change in strength over time due to the corrosion effect of category.

Table 3.6.4 Result after processing [118]

	Stochastic results									
time, [months]	0	17	27	46	65	86	100	120	136	143
ε, [mm]	8.760	9.034	9.939	9.120	5.490	6.125	5.554	5.295	2.654	2.274
σ, [MPa]	363.43	347.40	366.40	405.30	357.90	372.90	369.80	345.50	377.90	363.90
	Average results									
time, [months]	0	17	27	46	65	86	100	120	136	143
ε, [mm]	8.950	8.366	9.836	9.174	5.008	5.569	5.339	4.173	2.399	2.132
σ, [MPa]	359.19	343.70	365.90	376.90	352.90	358.60	379.90	316.10	295.70	339.40

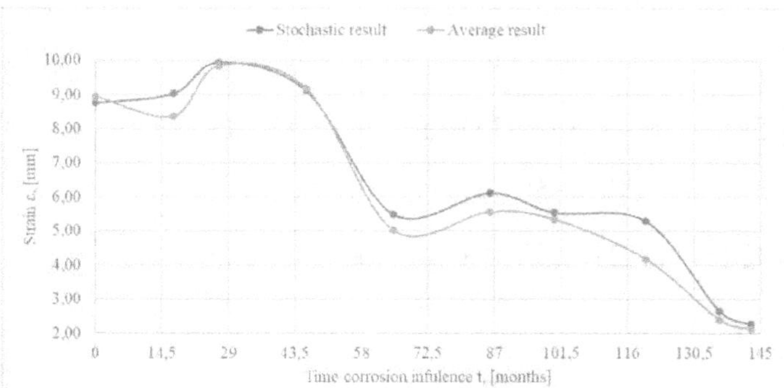

(a)

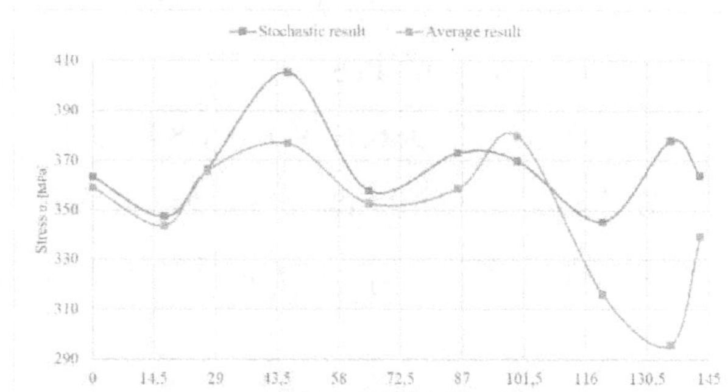

(b)

Figure 3.6.4 (a) Graph of the variation of the strain deformation depending on the corrosion effect over time; [118]; (b) Graph of the change of stress strength depending on the corrosion effect over time [118];

Using the polynomial approximation [14-15,114-119] from fig. 3.6.4, functional dependence for the strain and stress point in dependence of the time of the influence of the corrosion (in months) is established in functions or equations.

6.4.1. Stochastic functions

- Strain function [118]

$$\varepsilon(t) = 4.495855307765.10^{-15}.t^{9} - 2.97153032077.10^{-12}.t^{8} + 8.216476137477.10^{-10}.t^{7} - 1.2304694326356.10^{-7}.t^{6} + 1.0785385071753187.10^{-5}.t^{5} - 5.5740721964079448.10^{-4}.t^{4} + 1.6242605586171938.10^{-2}.t^{3} - 2.3962530738507931.10^{-1}.t^{2} + 1.38942800511922.t + 8.760 \quad (3.6.13)$$

- Stress function [118]

$$\sigma(t) = 5.18590474859.10^{-14}.t^{9} - 3.6946436134164.10^{-11}.t^{8} + 1.100020945672889.10^{-8}.t^{7} - 1.7725357694515535.10^{-6}.t^{6} + 1.6713806543659172.10^{-4}.t^{5} - 9.292042727450697.10^{-3}.t^{4} + 2.9047558639397697.10^{-1}.t^{3} - 4.5099081586850129.t^{2} + 25.736251427109956.t + 363.43 \quad (3.6.14)$$

6.4.2. Average functions

- Strain function [118]

$$\varepsilon(t) = 3.481457032279549.10^{-15}.t^{9} - 2.337153570162.10^{-12}.t^{8} + 6.5559274053.10^{-10}.t^{7} - 9.931988635863629.10^{-8}.t^{6} + 8.757240597872699.10^{-6}.t^{5} - 4.503505080145862.10^{-4}.t^{4} + 1.278207198422854.10^{-2}.t^{3} - 1.759323053282189.10^{-1}.t^{2} + 0.86976088608913771.t + 8.950 \quad (3.6.15)$$

- Stress function [118]

$$\sigma(t) = -3.059547202986.10^{-14}.t^{9} + 1.898537827165.10^{-11}.t^{8} - 4.882206148759711.10^{-8}.t^{7} + 6.7810733671137.10^{-7}.t^{6} - 5.5867022078063.10^{-5}.t^{5} + 2.8404859501200566.10^{-3}.t^{4} - 8.970433471157449.10^{-2}.t^{3} + 1.652346194143302.t^{2} - 13.21825571394335.t + 359.19 \quad (3.6.16)$$

6.5. Ultimate strength point

On table 3.6.5 is present a result after processing by stochastic method and average method. In fig. 3.6.5(a) shows the relationship between the change in relative strain versus the action time of the corrosion category, and in fig. 3.6.5(b) shows the relationship between the change in strength over time due to the corrosion effect of category.

Table 3.6.5 Result after processing [114]

	Stochastic results									
time, [months]	0	17	27	46	65	86	100	120	136	143
ε, [mm]	28.540	28.370	23.790	19.730	15.270	13.120	11.250	8.090	6.070	3.430
σ, [MPa]	460.91	448.90	450.90	474.80	444.00	446.90	449.90	413.90	433.90	451.90
	Average results									
time, [months]	0	17	27	46	65	86	100	120	136	143
ε, [mm]	29.600	23.410	23.440	19.440	12.470	12.250	12.000	8.890	4.550	3.550
σ, [MPa]	447.20	437.10	441.30	444.20	423.70	431.40	443.00	390.60	354.40	421.00

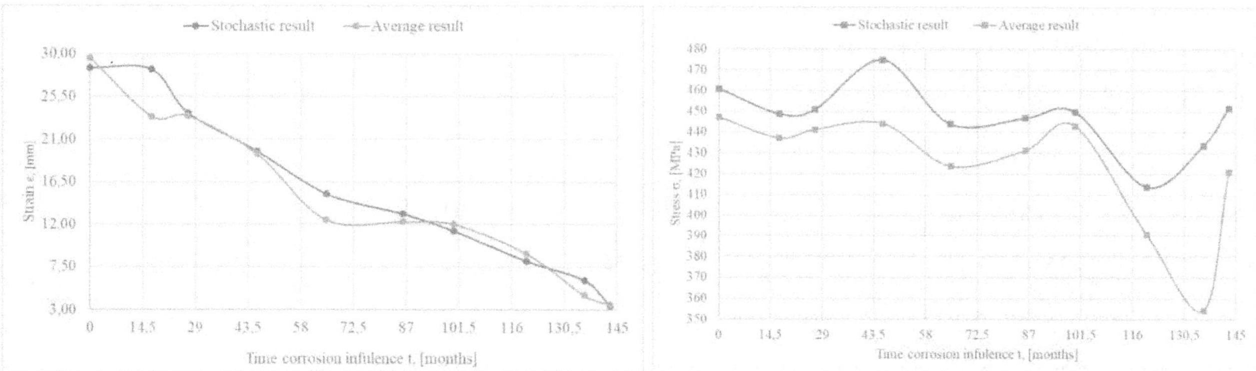

(a) (b)

Figure 3.6.5 (a) Graph of the variation of the strain deformation depending on the corrosion effect over time; [114]; (b) Graph of the change of stress strength depending on the corrosion effect over time [114];

Using the polynomial approximation [14-15,114-119] from fig. 3.6.5, functional dependence for the strain and stress point in dependence of the time of the influence of the corrosion (in months) is established in functions or equations.

6.5.1. Stochastic functions

- Strain function [114]

$$\varepsilon(t) = 2.5104382719.10^{-15}.t^{9} - 1.78942955261847.10^{-12}.t^{8} + 5.372342064887.10^{-10}.t^{7} - 8.830705223229639.10^{-8}.t^{6} + 8.646357701791169.10^{-6}.t^{5} \quad (3.6.17)$$

$- 5.1376086730459537.10^{-4}.t^4 + 1.8032936423633045.10^{-2}.t^3$

$- 3.4211856048264322.10^{-1}.t^2 + 2.50957930731836.t + 28.540$

- Stress function [114]

$$\sigma(t) = 1.5936909866.10^{-14}.t^9 - 1.26984960698977.10^{-11}.t^8 + 4.1762358074453579.10^{-9}.t^7 - 7.359138109345649.10^{-7}.t^6 + 7.52840101645.10^{-5}.t^5 - 4.5147821053993856.10^{-3}.t^4 + 1.5179779263840529.10^{-1}.t^3 - 2.5372356024528946.t^2 + 15.399449178934438.t + 460.91 \quad (3.6.18)$$

6.5.2. Average functions

- Strain function [114]

$$\varepsilon(t) = 3.25695513814.10^{-15}.t^9 - 2.1944555543722.10^{-12}.t^8 + 6.16889793230442.10^{-10}.t^7 - 9.33194876743229.10^{-8}.t^6 + 8.1551453874354.10^{-6}.t^5 - 4.0952483016803542.10^{-4}.t^4 + 1.0989665408898452.10^{-2}.t^3 - 1.300059567684525.10^{-1}.t^2 + 1.193226557958.10^{-1}.t + 29.600 \quad (3.6.19)$$

- Stress function [114]

$$\sigma(t) = 4.32397593589.10^{-16}.t^9 - 8.2520186695799.10^{-13}.t^8 + 4.307288364715.10^{-10}.t^7 - 1.0097022083256739.10^{-7}.t^6 + 1.229613913566859.10^{-5}.t^5 - 8.028009087845683.10^{-4}.t^4 + 2.68087105976477.10^{-2}.t^3 - 3.9133931372974617.10^{-1}.t^2 + 1.3614053185390875.t + 447.20 \quad (3.6.20)$$

6.6. Fracture point

On table 3.6.6 is present a result after processing by stochastic method and average method. In fig. 3.6.6(a) shows the relationship between the change in relative strain versus the action time of the corrosion category, and in fig. 3.6.6(b) shows the relationship between the change in strength over time due to the corrosion effect of category.

Using the polynomial approximation [14-15,114-119] from fig. 3.6.6, functional dependence for the strain and stress point in dependence of the time of the influence of the corrosion (in months) is established in functions or equations.

Table 3.6.6 Result after processing

	Stochastic results									
time, [months]	0	17	27	46	65	86	100	120	136	143
ε, [mm]	40.410	36.080	28.360	23.420	17.600	16.640	12.990	10.230	6.750	5.170
σ, [MPa]	286.41	282.70	284.40	297.40	288.90	285.90	305.90	289.90	287.90	322.90
	Average results									
time, [months]	0	17	27	46	65	86	100	120	136	143
ε, [mm]	38.800	29.180	28.250	23.020	15.120	14.370	14.160	10.280	5.630	4.310
σ, [MPa]	285.08	284.90	290.50	304.90	292.70	298.30	300.30	263.70	231.00	295.30

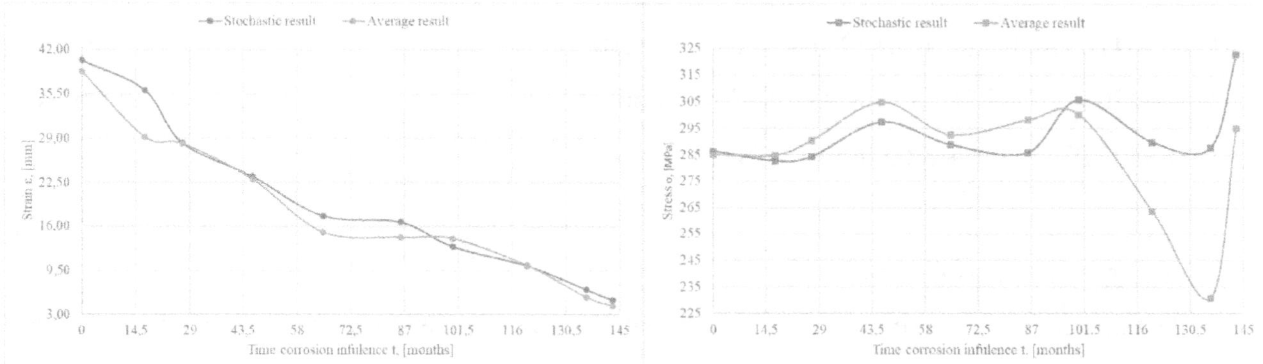

(a) (b)

Figure 3.6.6 (a) Graph of the variation of the strain deformation depending on the corrosion effect over time; (b) Graph of the change of stress strength depending on the corrosion effect over time;

6.6.1. Stochastic functions

- Strain function

$$\varepsilon(t) = 9.125677417.10^{-15}.t^9 - 6.1683752055289.10^{-12}.t^8 + 1.754525800773278.10^{-9}.t^7 - 2.7285537920010273.10^{-7}.t^6 + 2.5224002801027423.10^{-5}.t^5 - 1.4102112028989.10^{-3}.t^4 + 4.625089639493805.10^{-2}.t^3 - 8.0616185507777871.10^{-1}.t^2 + 5.2526881513512915.t + 40.410 \quad (3.6.21)$$

- Stress function

$$\sigma(t) = -2.068053489479.10^{-14}.t^9 + 1.28288023142.10^{-11}.t^8 \quad (3.6.22)$$

$$- 3.2603647032362688.10^{-9}.t^7 + 4.3771018928383663.10^{-7}.t^6$$
$$- 3.3424649909516717.10^{-5}.t^5 + 1.46238562067229.10^{-3}.t^4$$
$$- 3.554995837907235.10^{-2}.t^3 + 4.5598787018068077.t^2 - 2.637015122096777.t$$
$$+ 286.41$$

6.6.2. Average functions

- Strain function

$$\varepsilon (t) = 2.6767979831.10^{-15}.t^9 - 1.82921260762369.10^{-12}.t^8 + 5.2101206235758.10^{-10}.t^7$$
$$- 7.9676161443955173.10^{-8}.t^6 + 7.006867957618.10^{-6}.t^5$$
$$- 3.5076788918211058.10^{-4}.t^4 + 9.1654324308519681.10^{-3}.t^3 \qquad (3.6.23)$$
$$- 9.5680556571870717.10^{-2}.t^2 - 0.348740459359389.t + 38.800$$

- Stress function

$$\sigma (t) = 1.844096302757.10^{-14}.t^9 - 1.2359854614.10^{-11}.t^8 + 3.52061062278479.10^{-9}.t^7$$
$$- 5.5125530410524.10^{-7}.t^6 + 5.1201166736256556.10^{-5}.t^5$$
$$- 2.8369864263172967.10^{-3}.t^4 + 8.936559384236567.10^{-2}.t^3 \qquad (3.6.24)$$
$$- 1.417383829798629.t^2 + 8.6226590781821528.t + 285.08$$

6.7. Young's modulus (E-modulus)

On table 3.6.7 is present a result is presented after processing, as the given values of the E-modulus are calculated as the tangent of the angle that the deformation curve makes with the abscissa axis of the σ-ε diagram and subsequently (the obtained values) are processed using the stochastic method and average method. Fig. 3.3.7 shows the relationship between the change in E-modulus and the time of action of the corrosion category.

Using the polynomial approximation [14-15,114-119] from fig. 3.3.7, functional dependence for the E-modulus in dependence of the time of the influence of the corrosion (in months) is established in functions or equations.

6.7.1. Stochastic functions

$$E (t) = 3.140435153.10^{-14}.t^9 - 1.978764074846.10^{-11}.t^8 + 5.1986575362499735.10^{-9}.t^7$$
$$- 7.393096415214466.10^{-7}.t^6 + 6.1883903215233037.10^{-5}.t^5$$
$$- 3.104082619691802.10^{-3}.t^4 + 9.0200870385961335.10^{-2}.t^3 - 1.329503608663038.t^2 \qquad (3.6.25)$$
$$+ 4.9944107074842865.t + 211.50$$

6.7.2. Average functions

$$E(t) = 3.995904562333.10^{-14}.t^{9} - 2.49162299145699.10^{-11}.t^{8} + 6.421992424733.10^{-9}.t^{7} - 8.8450019660779583.10^{-7}.t^{6} + 7.0356273305277744.10^{-5}.t^{5} - 3.267010332252685.10^{-3}.t^{4} + 8.5326573634903966.10^{-2}.t^{3} - 1.1138584117448611.t^{2} + 3.9792034744765701.t + 193.10 \quad (3.6.26)$$

Table 3.6.7 Result after processing

	Stochastic results									
time, [months]	0	17	27	46	65	86	100	120	136	143
E, [GPa]	211.50	168.10	153.50	148.70	156.20	151.90	131.60	153.50	184.50	232.00
	Average results									
time, [months]	0	17	27	46	65	86	100	120	136	143
E, [GPa]	193.10	166.20	159.10	143.60	156.30	164.40	136.50	157.50	181.20	206.60

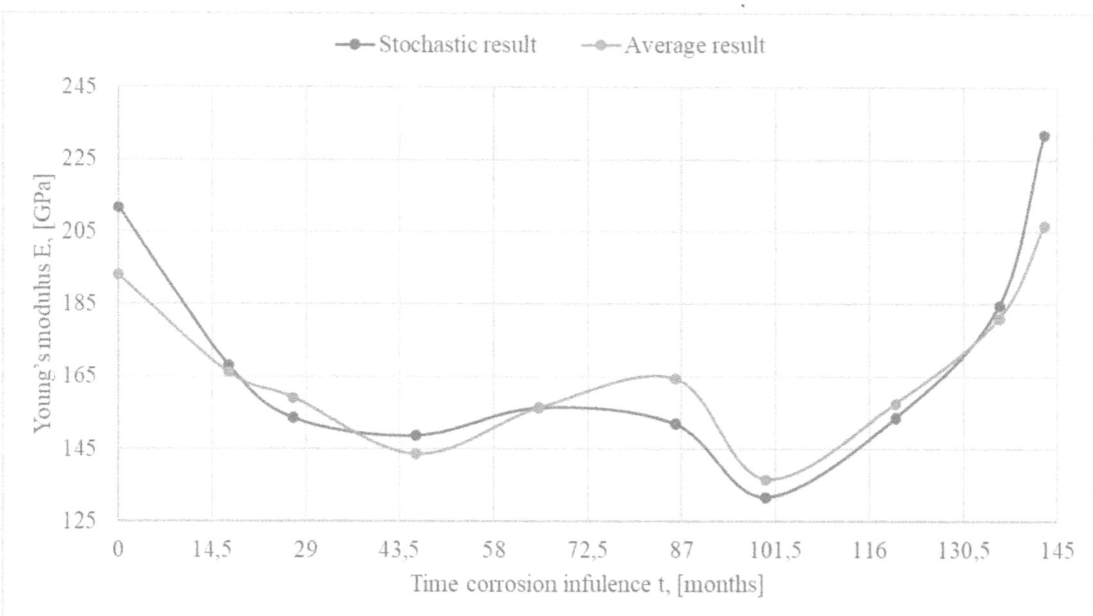

Figure 3.6.7 Graph of the variation of the Young's modulus (E-modulus) depending on the corrosion effect over time;

7. Summary

From the established equations (eq. 3.1.1 - eq. 3.6.26), it is found that the trend in the mechanical properties of steel as a result of corrosion is a function of a 9th degree polynomial, which can be summarized as:

- Strain function

$$\varepsilon\ (t) = A_1.t^9 + A_2.t^8 + A_3.t^7 + A_4.t^6 + A_5.t^5 + A_6.t^4 + A_7.t^3 + A_8.t^2 + A_9.t + \varepsilon_i$$

where: A_1, A_2, A_3, A_4, A_5, A_6, A_7, A_8 и A_9 are coefficients.

- Stress function

$$\sigma\ (t) = B_1.t^9 + B_2.t^8 + B_3.t^7 + B_4.t^6 + B_5.t^5 + B_6.t^4 + B_7.t^3 + B_8.t^2 + B_9.t + \sigma_i$$

where: B_1, B_2, B_3, B_4, B_5, B_6, B_7, B_8 и B_9 are coefficients.

- Young's modulus (E-modulus)

$$E(t) = C_1.t^9 + C_2.t^8 + C_3.t^7 + C_4.t^6 + C_5.t^5 + C_6.t^4 + C_7.t^3 + C_8.t^2 + C_9.t + E_i$$

where: C_1, C_2, C_3, C_4, C_5, C_6, C_7, C_8 и C_9 are coefficients.

CHAPTER IV. CONCLUSION

1. Scientific results

- Polynomial functions have been collected and systematized to determine the change in the main characteristics of the mechanical properties of corroded steel.
- Polynomial functions have been established to determine the trend of the main mechanical features of the stress-strain curve depending on the time of corrosion exposure.
- Three of the main mechanical properties have been established for the first time.

2. Practical results

- The established polynomial equations can be used in practice to verify and/or calculate the load-bearing capacity of a corroded steel structural element.
- Using a the established polynomial equations, the time interval in which the corroded steel element will lose its load-bearing capacity can be predicted with sufficient accuracy for practice.
- Using the established polynomial functions, the exact moment in time when the corroded steel load-bearing element will lose its load-bearing capacity in the corresponding corrosion category can be calculated.

3. Applied results

- The established polynomial functions can be used in determining the degree of suitability of a given structural facility with corrosion.
- With the established polynomial equations, the degree of suitability of a structural element with corrosion can be determined depending on the corrosion category and its time impact.
- It is possible to predict the time period in which a structure with corrosion will fall into an emergency, pre-emergency or dangerous state (unfit for use).
- The time during which a structure with corrosion can be used for its intended purpose safely can be calculated.
- It is possible to predict the point in time when maintenance repair activities need to be carried out.

4. Main conclusion

Polynomial functions that determine the change in mechanical properties of corroded steel over time are necessary to prevent accidents caused by corrosion.

ANNEX № 1

Experimental data of corroded steel test specimens (tubes, rods)

1. First experimental data set

Table 5.1.1 Time of corrosion influence according to corrosion category

Corrosion category C1 [months]	Corrosion category C2 [months]	Corrosion category C3 [months]	Corrosion category C4 [months]	Corrosion category C5 [months]	Corrosion category CX [months]
9139	475	238	149	59	17

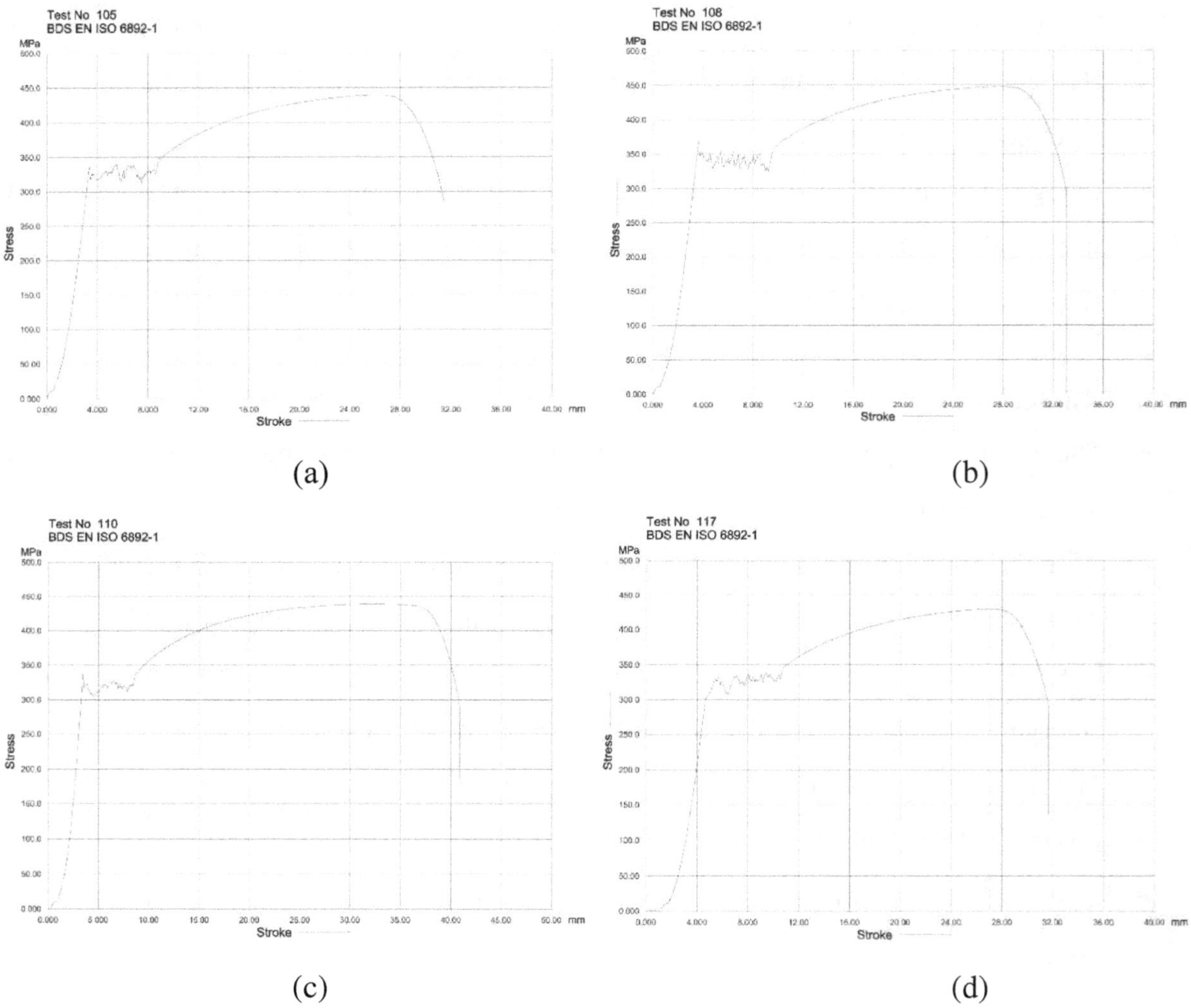

Figure 5.1.1 Stress-strain curve on corroded steel test specimens; (a) Test № 105 [120]; (b) Test № 108 [120]; (c) Test № 110 [120]; (d) Test № 117 [120];

2. Second experimental data set

Table 5.2.1 Time of corrosion influence according to corrosion category

Corrosion category C1 [months]	Corrosion category C2 [months]	Corrosion category C3 [months]	Corrosion category C4 [months]	Corrosion category C5 [months]	Corrosion category CX [months]
14769	768	384	240	96	28

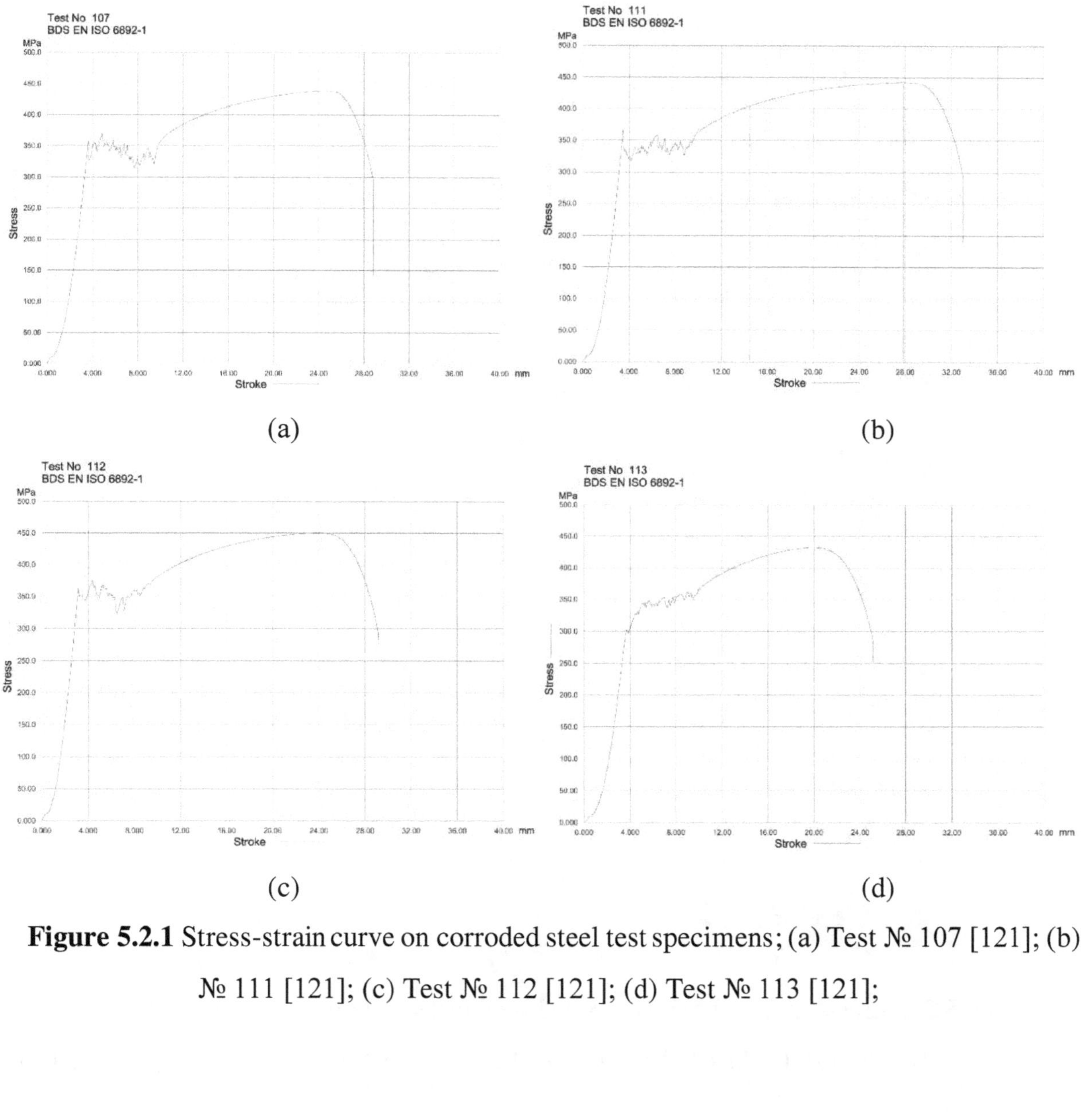

(a) (b) (c) (d)

Figure 5.2.1 Stress-strain curve on corroded steel test specimens; (a) Test № 107 [121]; (b) Test № 111 [121]; (c) Test № 112 [121]; (d) Test № 113 [121];

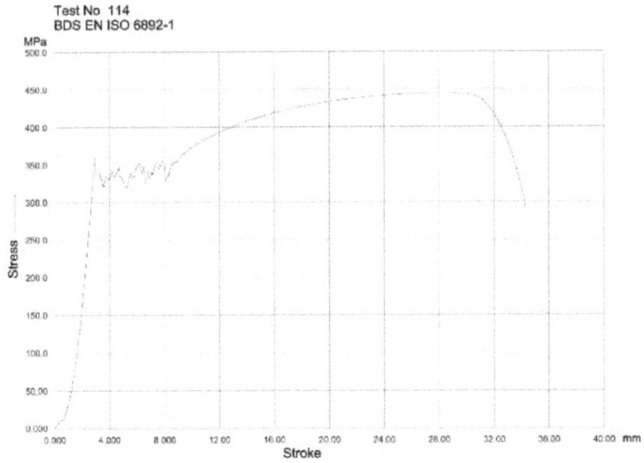

(a)

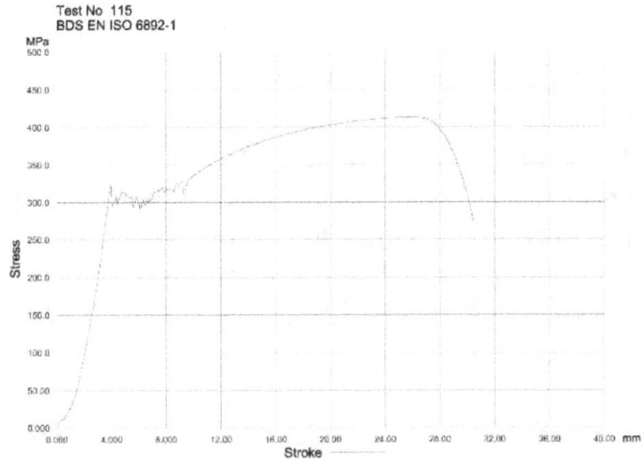

(b)

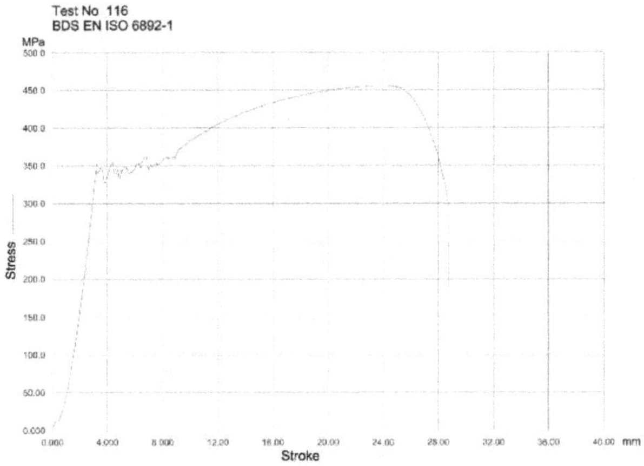

(c)

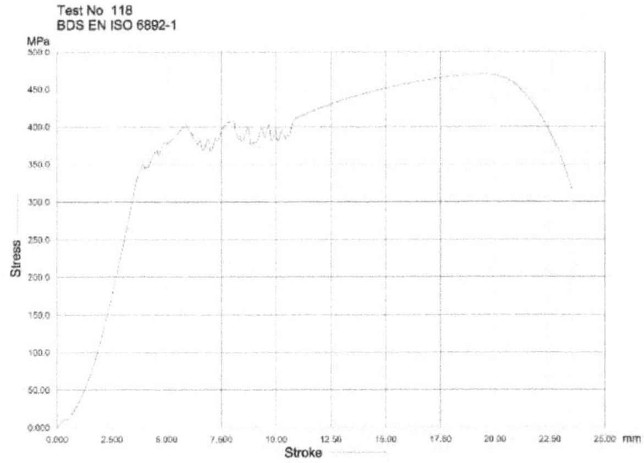

(d)

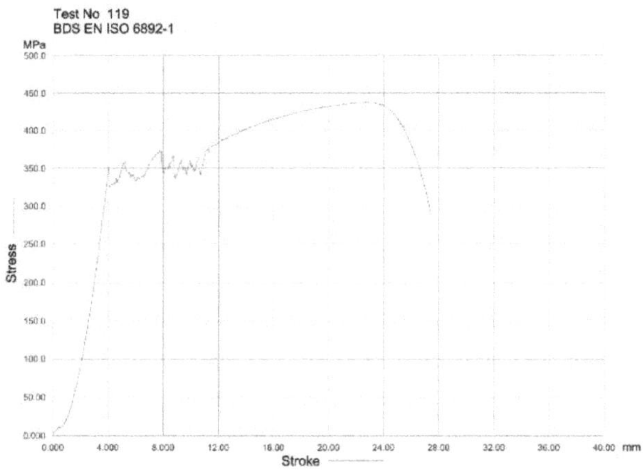

(e)

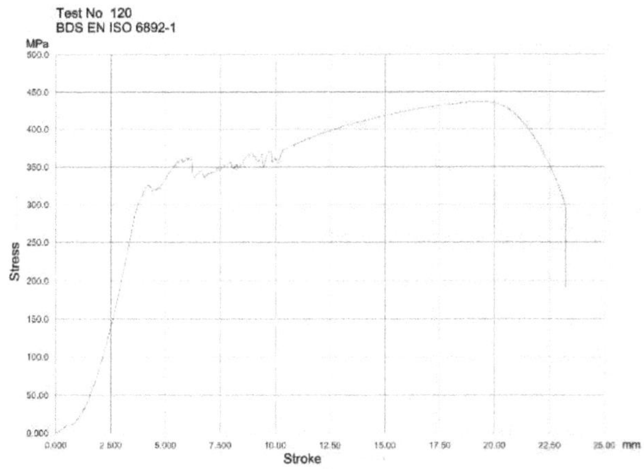

(f)

Figure 5.2.2 Stress-strain curve on corroded steel test specimens; (a) Test № 114 [121]; (b) Test № 115 [121]; (c) Test № 116 [121]; (d) Test № 118 [121]; (e) Test № 119 [121]; (f) Test № 120 [121];

3. Third experimental data set

Table 5.3.1 Time of corrosion influence according to corrosion category

Corrosion category C1 [months]	Corrosion category C2 [months]	Corrosion category C3 [months]	Corrosion category C4 [months]	Corrosion category C5 [months]	Corrosion category CX [months]
24923	1296	648	405	162	46

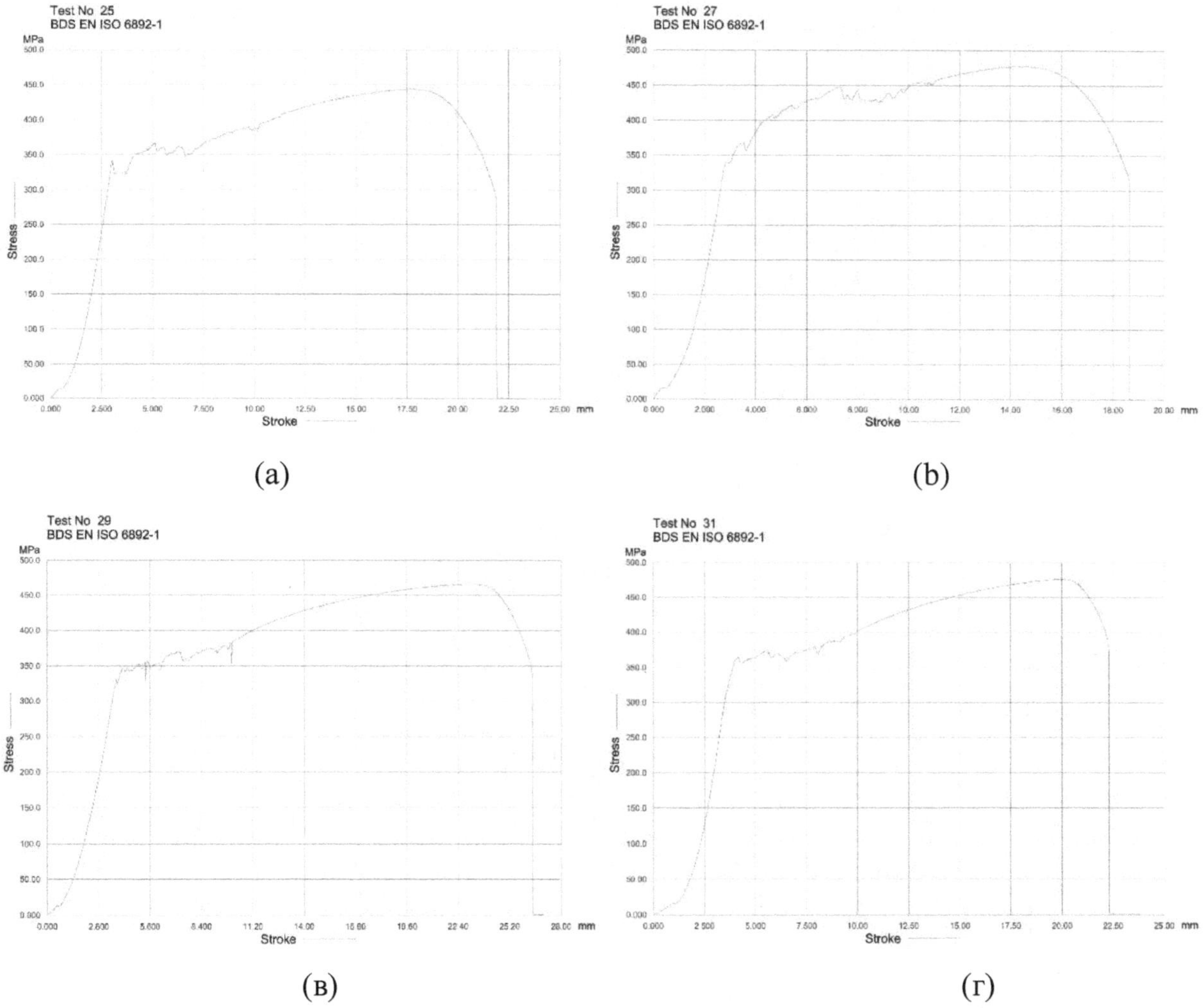

Figure 5.3.1 Stress-strain curve on corroded steel test specimens; (a) Test № 25 [122]; (b) Test № 27 [122]; (в) Test № 29 [122]; (г) Test № 31 [122];

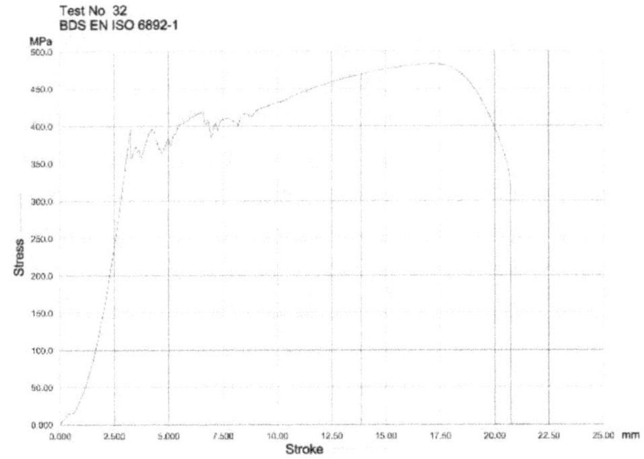

(a)

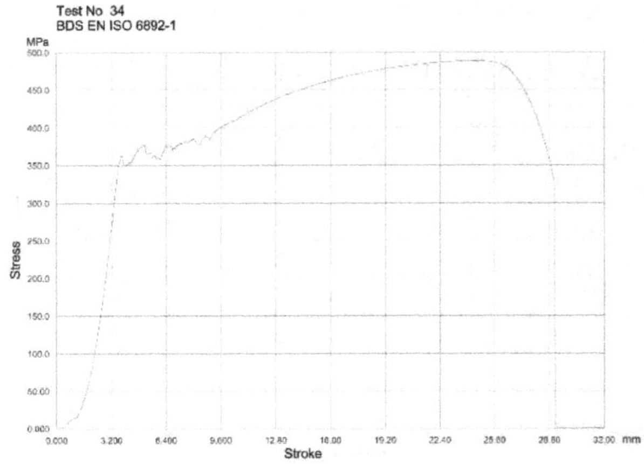

(b)

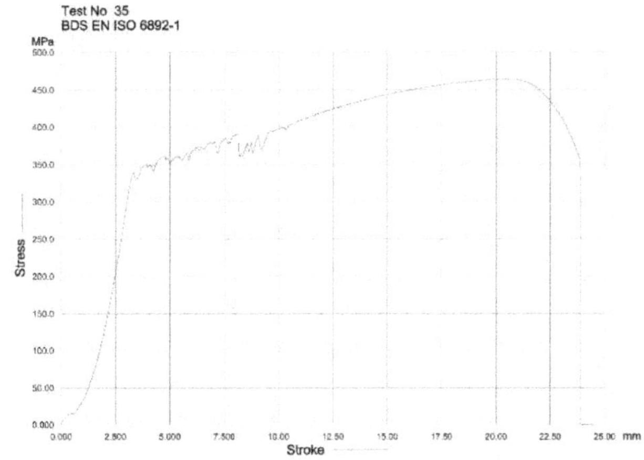

(c)

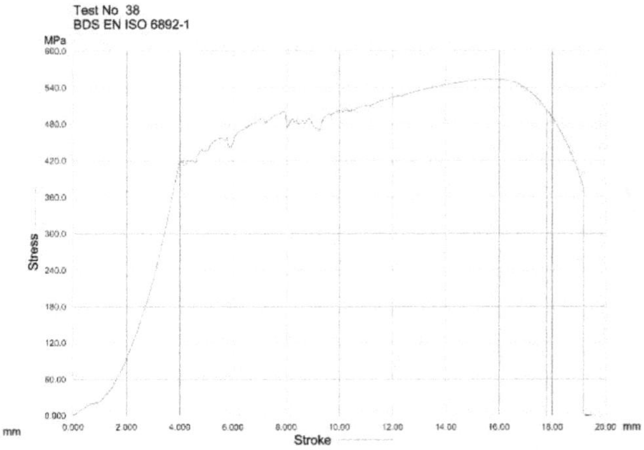

(d)

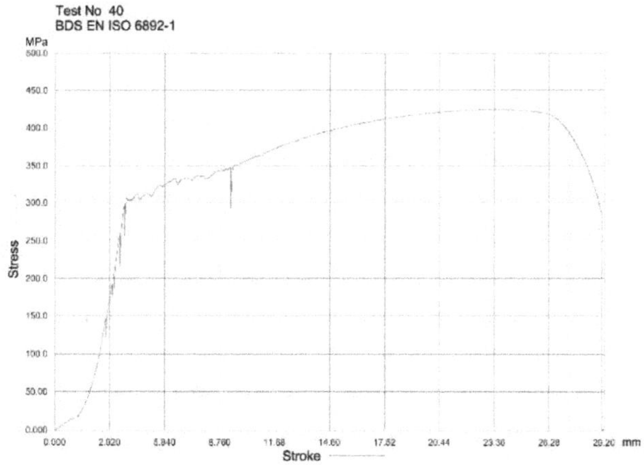

(e)

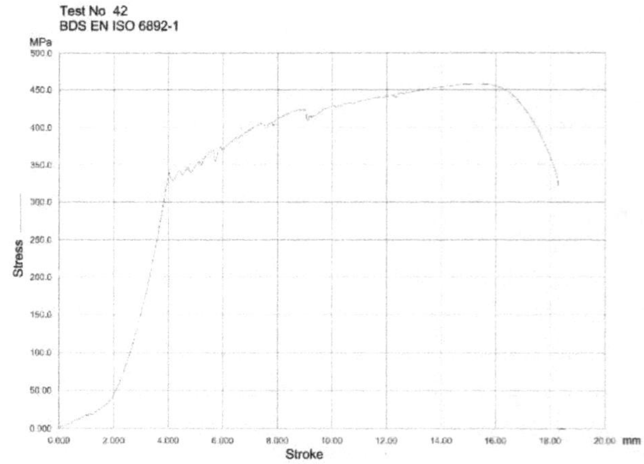

(f)

Figure 5.3.2 Stress-strain curve on corroded steel test specimens; (a) Test № 32 [122]; (b) Test № 34 [122]; (c) Test № 35 [122]; (d) Test № 38 [122]; (e) Test № 40 [122]; (f) Test № 42 [122];

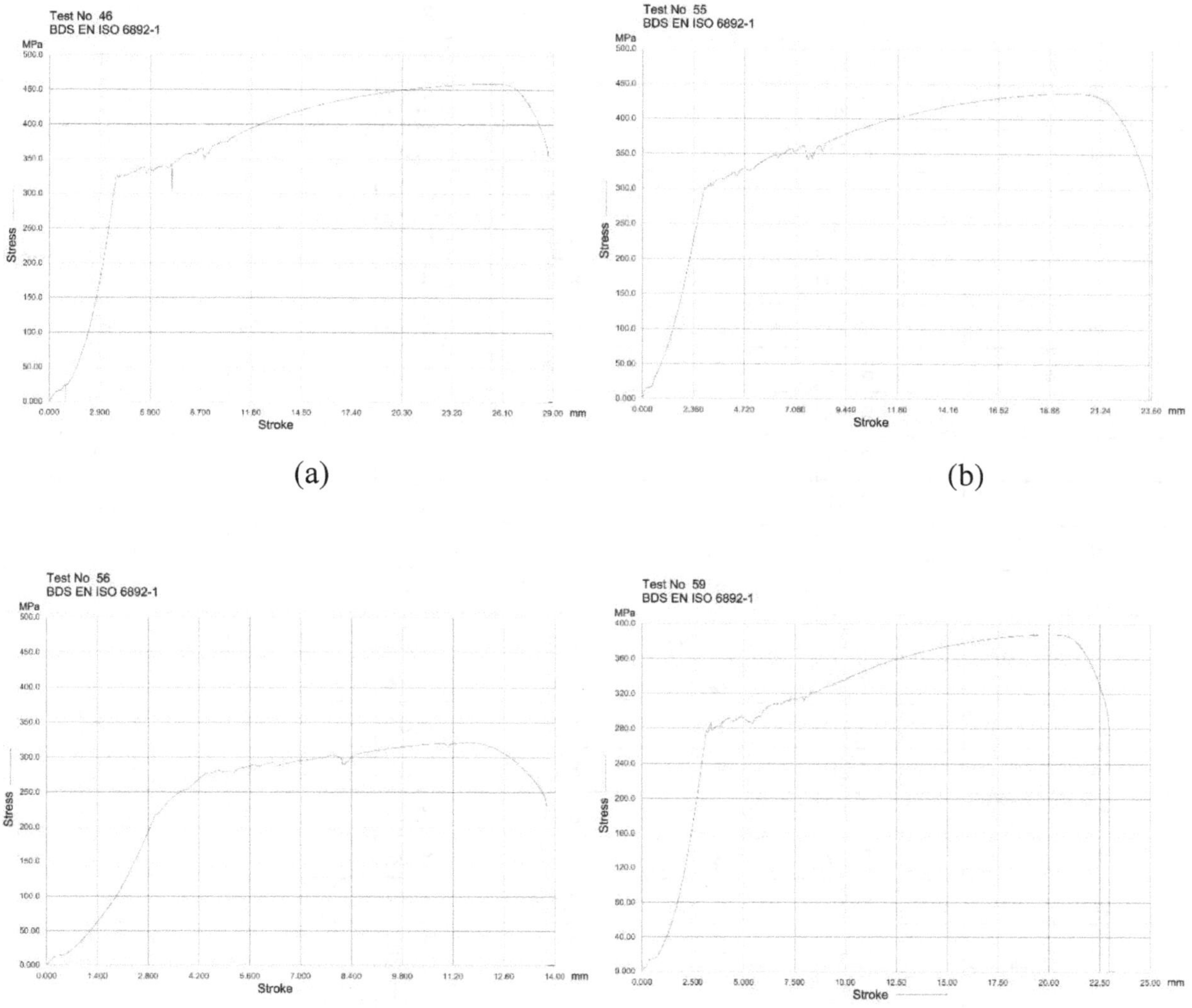

(a) (b) (c) (d)

Figure 5.3.3 Stress-strain curve on corroded steel test specimens; (a) Test № 46 [122]; (b) Test № 55 [122]; (c) Test № 56 [122]; (d) Test № 59 [122];

Table 5.3.2 Experimental data for change in diameter (in mm) along the length of the element according to the time of corrosion influence [129];

Points	Test specimens №						
	27	25	35	34	32	31	29
1	7.81	8.68	9.09	8.84	8.35	8.66	8.09
2	7.12	7.88	7.80	8.93	7.81	8.54	8.23
4	7.27	7.84	7.59	8.91	7.78	8.57	8.69
5	7.30	7.77	8.12	9.79	7.82	8.61	8.93
6	7.50	8.12	8.58	8.91	7.84	8.50	8.21

7	8.10	8.67	8.19	8.63	7.84	8.43	8.18
8	8.20	8.69	8.10	8.49	7.62	8.60	8.12
9	7.96	8.85	7.99	8.66	7.51	8.61	8.47
10	8.41	8.67	8.26	8.69	7.58	8.40	9.05
11	8.73	8.48	8.50	8.57	7.39	8.43	9.27
12	9.01	8.68	8.34	8.72	7.60	7.97	9.68
13	8.59	8.69	8.24	8.45	8.14	7.46	9.47
14	9.17	8.63	8.31	8.48	8.39	7.46	9.70
15	9.08	8.75	7.79	8.44	8.03	7.72	9.88
16	9.16	8.48	7.73	8.51	8.01	8.18	10.00
17	8.93	8.60	7.54	7.87	7.36	7.54	9.99
18	7.91	8.54	7.56	8.22	7.48	7.50	9.63
19	7.78	8.36	7.61	8.21	8.10	7.64	9.41
20	8.13	8.37	7.91	7.60	8.69	7.76	9.06
21	8.35	8.24	7.95	7.93	8.88	7.79	8.99
22	8.26	8.67	7.80	7.88	9.61	8.35	8.98
23	8.46	8.52	8.57	8.09	8.93	8.36	8.88
24	8.20	9.13	8.65	8.25	8.33	8.16	8.90
25	8.02	9.13	8.78	8.18	8.49	8.20	8.70
26	8.17	8.62	9.29	8.08	8.95	8.65	9.08
27	8.22	8.65	8.68	7.63	8.57	8.22	8.80
28	8.31	8.49	8.42	7.95	9.14	8.06	8.73
29	8.29	8.61	8.65	7.76	8.95	8.07	8.74
30	8.66	8.73	9.31	9.18	8.88	7.43	9.17

Table 5.3.3 Experimental data for change in diameter (in mm) along the length of the element according to the time of corrosion influence [129];

Points	Test specimens №						
	59	56	55	46	38	40	42
1	8.75	9.50	9.10	8.99	9.32	9.68	10.00
2	8.50	9.09	8.70	8.85	8.13	9.38	9.31
4	8.77	9.32	8.79	8.86	7.89	8.64	9.91
5	8.27	8.73	7.70	9.19	7.54	9.11	9.52
6	8.30	10.00	7.81	9.05	7.32	9.28	8.18
7	8.80	10.00	7.57	8.54	7.21	8.02	8.55
8	9.26	9.80	7.43	8.56	7.82	7.86	8.80
9	8.84	9.20	7.57	8.01	7.69	7.45	9.01

10	9.04	7.83	7.36	8.26	7.90	7.99	9.58
11	7.82	8.21	7.65	8.45	8.48	8.01	9.42
12	7.53	8.03	7.31	7.94	9.00	7.65	8.69
13	7.61	8.88	7.29	7.37	8.75	7.96	8.67
14	8.75	9.03	7.69	7.58	8.35	8.58	8.39
15	8.95	9.09	7.68	7.73	8.16	8.38	8.05
16	8.78	8.92	8.10	8.13	7.79	8.00	7.11
17	9.40	8.83	8.13	8.92	7.55	8.38	7.87
18	8.30	8.75	8.35	9.24	7.31	8.76	8.08
19	7.90	9.23	8.69	9.24	7.48	9.02	8.27
20	9.56	9.03	8.67	9.43	8.16	8.71	8.92
21	9.00	8.93	8.88	8.93	8.12	8.64	8.94
22	8.83	8.23	8.10	8.52	7.55	7.61	9.41
23	9.31	8.71	8.30	7.76	8.10	8.61	9.14
24	8.91	9.06	8.88	8.76	8.02	8.70	8.68
25	8.82	8.86	8.51	8.26	8.08	8.52	8.93
26	9.35	8.19	8.13	8.57	8.78	8.44	8.52
27	9.68	8.28	8.04	8.62	8.78	8.53	8.37
28	10.00	8.82	8.40	8.44	8.90	9.06	8.35
29	10.00	8.10	8.99	8.50	8.74	8.27	8.19
30	8.43	9.69	8.86	8.42	8.83	7.94	8.00

4. Fourth experimental data set

Table 5.4.1 Time of corrosion influence according to corrosion category

Corrosion category C1 [months]	Corrosion category C2 [months]	Corrosion category C3 [months]	Corrosion category C4 [months]	Corrosion category C5 [months]	Corrosion category CX [months]
34892	1814	907	567	227	65

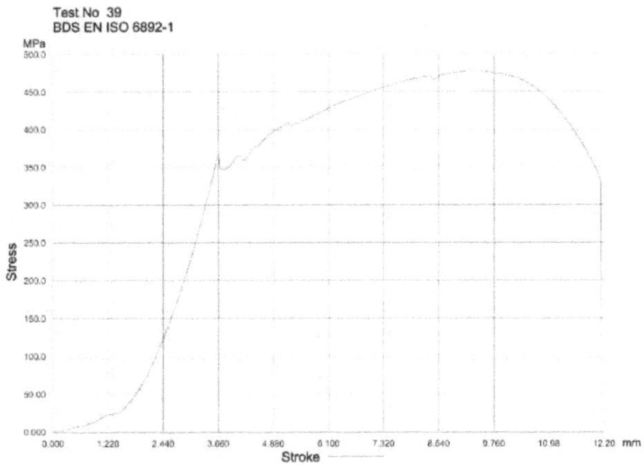

(a)

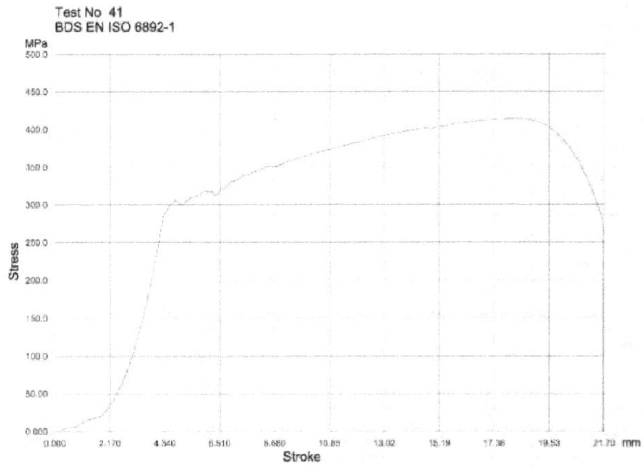

(b)

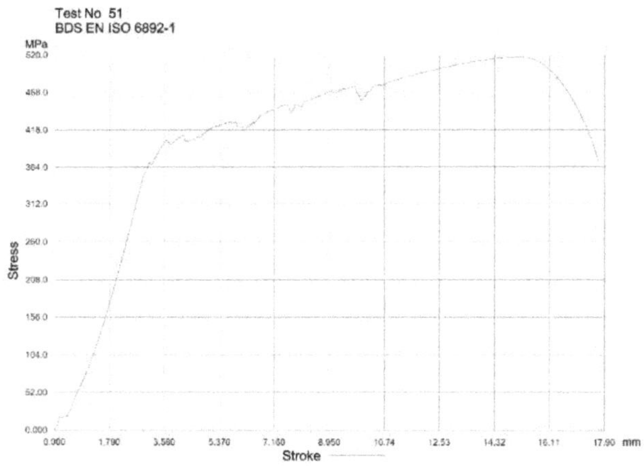

(c)

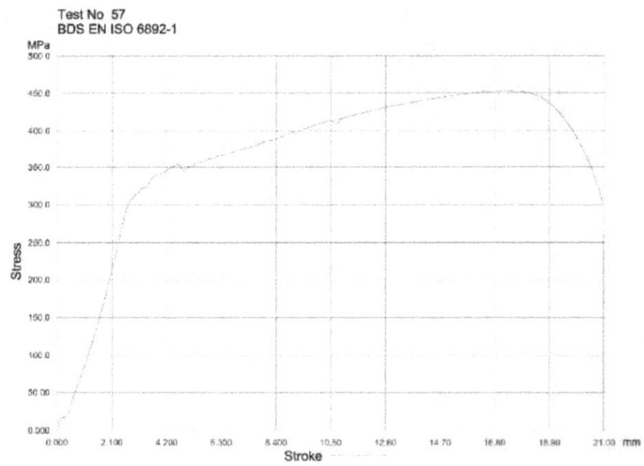

(d)

Figure 5.4.1 Stress-strain curve on corroded steel test specimens; (a) Test № 39 [123]; (b) Test № 41 [123]; (c) Test № 51 [123]; (d) Test № 57 [123];

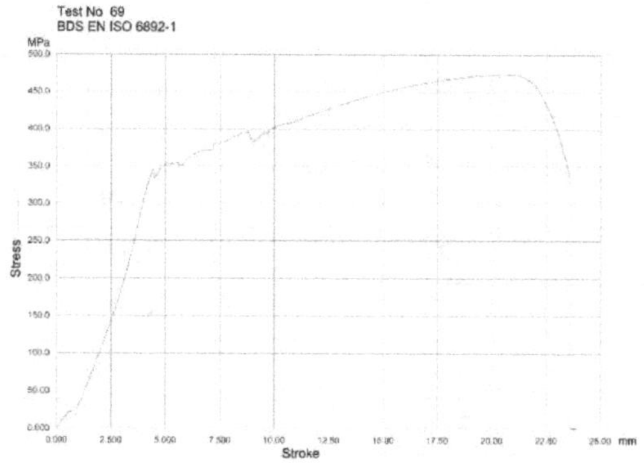

Figure 5.4.2 Stress-strain curve on corroded steel test specimens - Test № 69 [123];

Table 5.4.2 Experimental data for change in diameter (in mm) along the length of the element according to the time of corrosion influence [130];

Points	Test specimens №			
	69	57	51	41
1	7.96	9.46	9.95	7.66
2	7.49	9.40	9.53	7.52
4	7.10	9.00	9.87	7.58
5	6.40	9.40	9.58	7.01
6	6.55	9.52	9.75	7.54
7	6.46	9.48	8.99	6.60
8	5.98	9.31	8.98	6.48
9	5.94	9.10	8.38	6.82
10	6.12	9.38	8.10	6.89
11	7.09	8.78	8.06	6.81
12	7.17	8.40	7.69	7.40
13	6.67	7.92	7.82	7.30
14	6.27	8.09	6.74	7.39
15	6.59	8.13	7.29	8.45
16	6.83	8.62	8.62	7.84
17	7.58	8.40	8.64	7.01
18	7.63	8.09	8.66	7.19
19	7.70	7.94	9.13	7.63
20	7.66	8.19	8.15	7.31
21	8.26	7.72	8.97	6.98
22	7.85	7.42	8.85	6.86

23	7.47	6.99	9.11	7.27
24	7.24	6.93	9.08	7.39
25	7.78	8.27	8.30	8.45
26	6.37	8.89	8.64	8.82
27	6.52	9.14	8.96	8.17
28	6.17	8.96	8.37	7.36
29	6.38	8.18	8.46	7.74
30	6.53	7.97	9.00	7.56

5. Fifth experimental data set

Table 5.5.1 Time of corrosion influence according to corrosion category

Corrosion category C1 [months]	Corrosion category C2 [months]	Corrosion category C3 [months]	Corrosion category C4 [months]	Corrosion category C5 [months]	Corrosion category CX [months]
46154	2400	1200	750	300	86

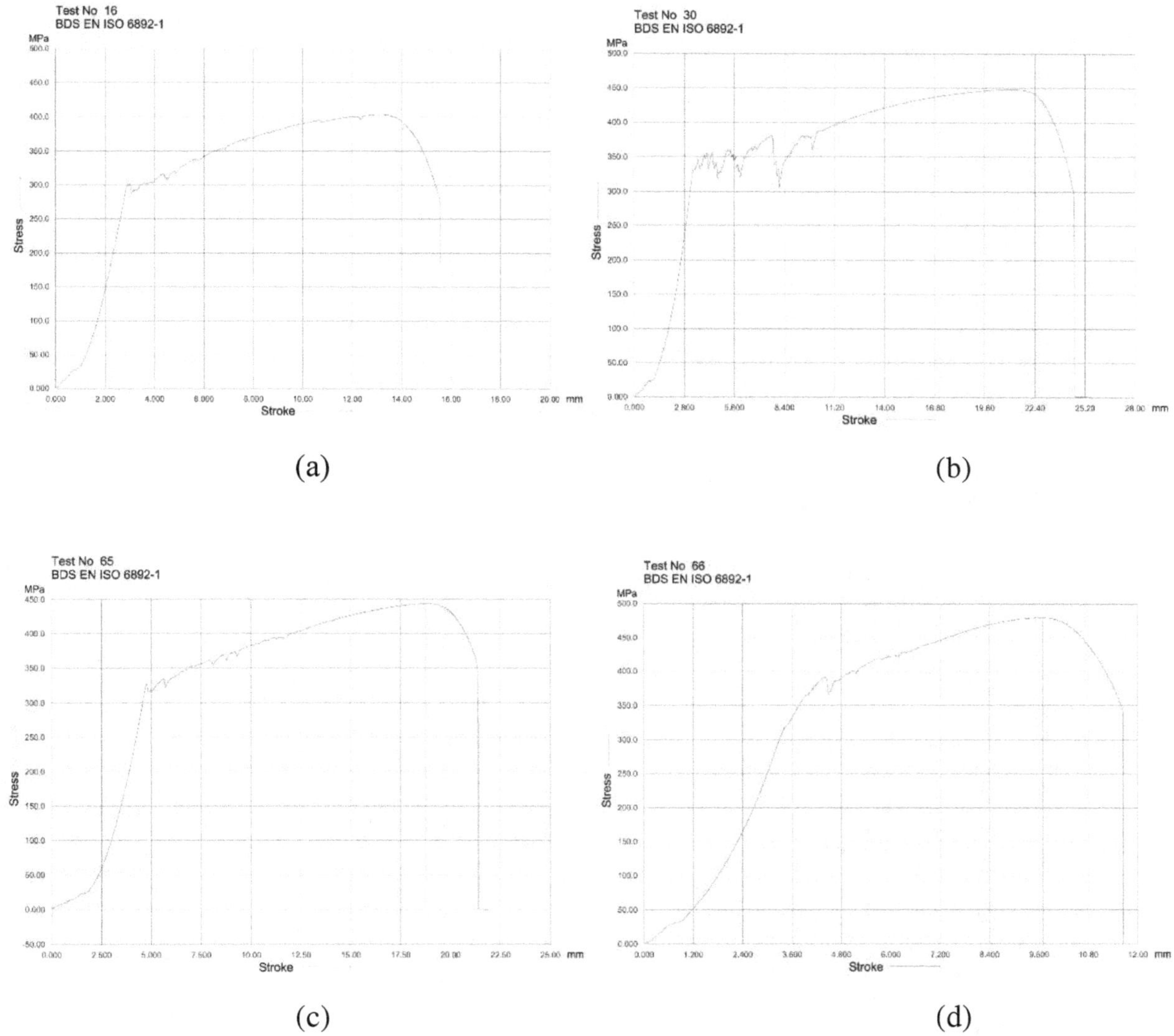

Figure 5.5.1 Stress-strain curve on corroded steel test specimens; (a) Test № 16 [124]; (b) Test № 30 [124]; (c) Test № 65 [124]; (d) Test № 66 [124];

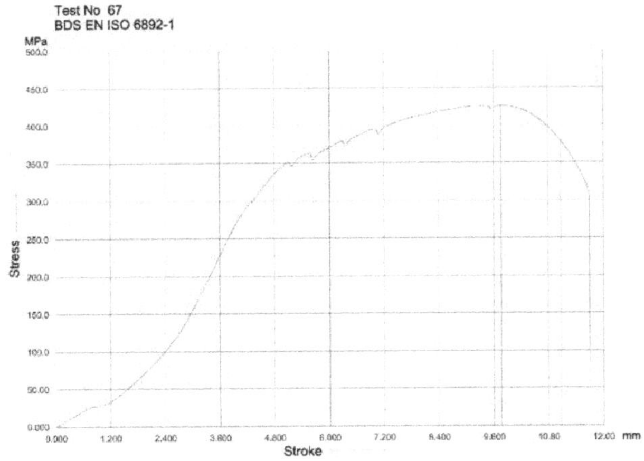

(a)

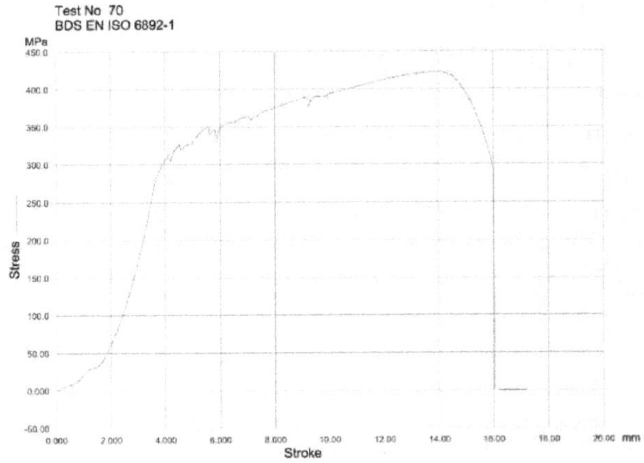

(b)

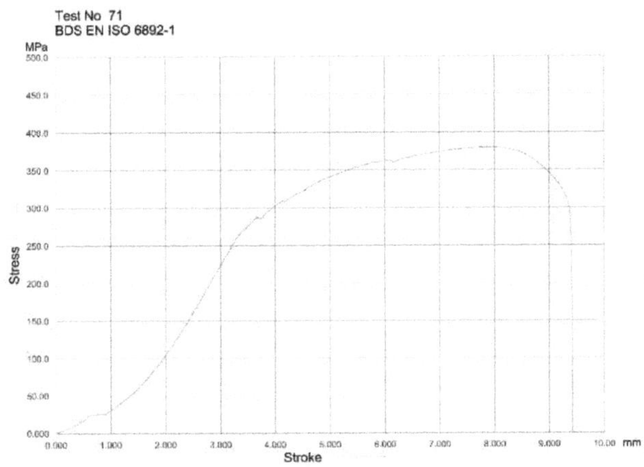

(c)

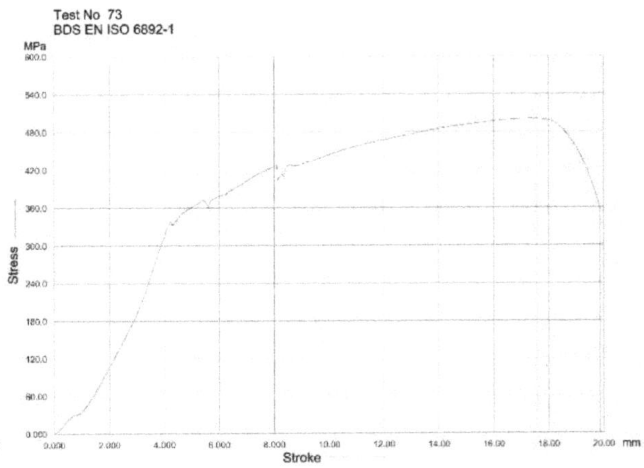

(d)

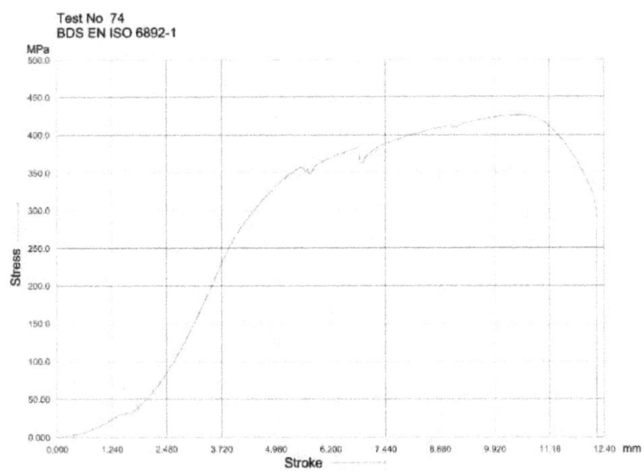

(e)

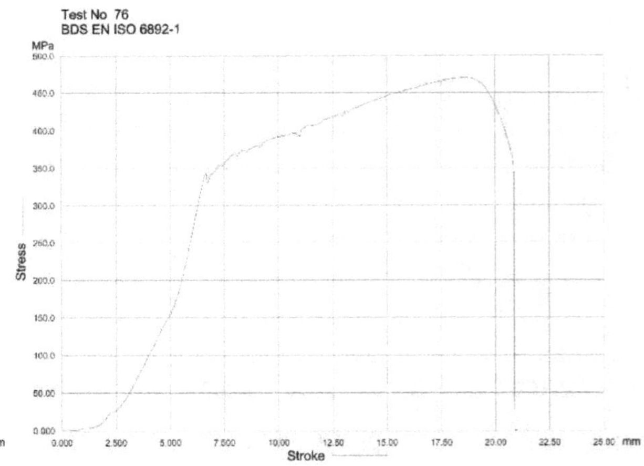

(f)

Figure 5.5.2 Stress-strain curve on corroded steel test specimens; (a) Test № 67 [124]; (b) Test № 70 [124]; (c) Test № 71 [124]; (d) Test № 73 [124]; (e) Test № 74 [124]; (f) Test № 78 [124];

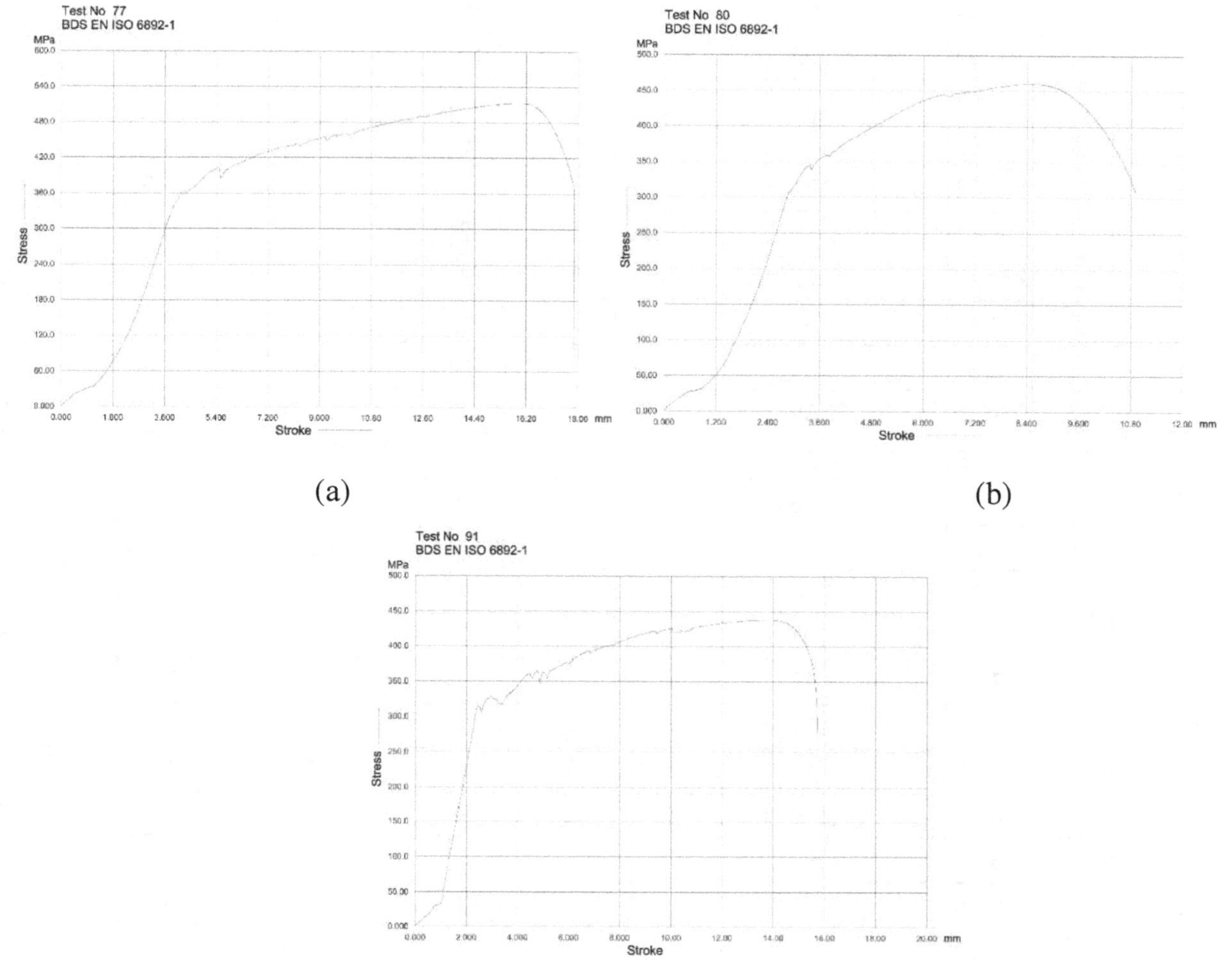

(c)

Figure 5.5.2 Stress-strain curve on corroded steel test specimens; (a) Test № 77 [124]; (b) Test № 80 [124]; (c) Test № 91 [124]

Table 5.5.2 Experimental data for change in diameter (in mm) along the length of the element according to the time of corrosion influence [131];

Points	Test specimens №						
	16	91	30	80	76	77	73
1	6.49	6.24	7.96	7.13	7.09	7.79	8.89
2	5.85	5.36	6.39	6.88	6.12	6.40	8.16
4	5.40	6.49	6.30	7.47	6.47	5.84	7.70
5	5.59	6.23	6.39	6.45	7.30	6.51	7.17
6	6.33	6.41	6.21	5.60	7.51	6.06	8.30

7	6.17	7.54	6.21	5.33	7.24	6.00	7.61
8	6.38	7.67	6.07	5.45	6.06	6.12	6.30
9	6.62	7.02	6.30	6.18	5.82	6.32	6.59
10	6.42	7.16	5.95	6.68	7.17	6.49	6.98
11	6.80	7.12	5.91	5.99	7.47	6.50	6.71
12	5.62	6.82	5.88	6.31	7.42	6.91	6.38
13	6.62	6.69	6.26	6.85	7.19	6.55	5.54
14	7.26	6.77	6.51	7.73	7.83	6.02	5.75
15	7.48	6.66	6.13	6.90	6.63	6.72	5.36
16	7.60	7.06	6.06	6.90	6.53	5.80	6.12
17	7.77	6.85	5.95	7.37	6.60	5.29	5.76
18	7.73	7.11	6.04	7.87	6.10	7.01	5.62
19	8.51	6.16	6.03	6.99	5.80	6.45	5.65
20	7.66	6.81	6.46	8.13	6.34	5.40	5.73
21	8.25	6.26	7.06	8.84	6.74	5.36	5.43
22	7.72	6.07	6.84	7.87	7.01	6.17	5.60
23	7.50	6.43	6.69	8.10	6.56	5.30	6.45
24	8.17	6.98	6.94	7.31	6.45	6.17	5.87
25	7.49	6.41	6.52	7.41	6.72	6.20	5.23
26	6.42	6.44	6.06	8.03	7.10	5.39	5.66
27	6.78	6.68	5.95	7.37	7.38	5.42	5.65
28	6.99	6.63	5.89	7.53	8.13	5.57	5.07
29	6.17	7.80	6.16	8.19	8.33	5.13	5.43
30	6.63	7.63	6.41	8.92	7.42	6.12	5.47

Table 5.5.3 Experimental data for change in diameter (in mm) along the length of the element according to the time of corrosion influence [131];

Points	Test specimens №					
	71	68	67	66	65	70
1	7.91	8.05	7.83	6.11	6.38	7.00
2	6.34	7.75	8.28	6.50	5.89	6.88
4	6.65	7.35	7.62	5.94	6.52	6.00
5	6.44	5.94	6.56	6.38	6.30	5.87
6	6.38	5.45	7.30	6.31	6.00	6.84
7	7.34	5.24	8.38	6.05	5.79	5.56
8	7.27	5.98	8.08	6.20	6.17	5.47
9	7.15	6.78	7.73	5.08	6.98	5.55

10	7.40	6.53	7.54	5.75	5.84	5.48
11	5.91	6.75	7.29	5.64	5.76	6.53
12	6.60	6.87	7.34	6.04	6.66	5.38
13	8.62	6.47	7.37	6.92	6.49	5.53
14	8.31	6.60	7.09	6.48	6.77	5.74
15	7.40	6.43	7.15	6.71	6.76	6.05
16	7.65	6.33	6.83	7.74	6.98	6.20
17	6.62	5.72	6.68	7.88	6.90	6.47
18	7.00	6.89	5.67	6.88	6.53	5.88
19	6.28	6.28	5.93	6.58	7.10	5.25
20	5.69	5.83	6.07	6.58	7.05	6.07
21	5.85	5.23	5.35	6.90	8.00	6.23
22	6.96	5.42	6.48	6.93	6.90	6.81
23	6.57	5.75	7.27	6.48	7.13	6.77
24	6.64	6.15	6.84	7.02	6.71	6.52
25	6.37	6.23	6.28	7.41	6.71	7.00
26	6.41	5.87	6.81	7.46	6.49	5.53
27	6.39	6.80	6.73	6.63	6.85	5.67
28	6.15	6.24	7.04	6.72	6.43	6.59
29	6.38	6.10	6.91	6.00	6.21	6.30
30	7.00	7.50	6.50	6.24	6.94	6.12

6. Sixth experimental data set

Table 5.6.1 Time of corrosion influence according to corrosion category

Corrosion category C1 [months]	Corrosion category C2 [months]	Corrosion category C3 [months]	Corrosion category C4 [months]	Corrosion category C5 [months]	Corrosion category CX [months]
54000	2808	1404	878	351	100

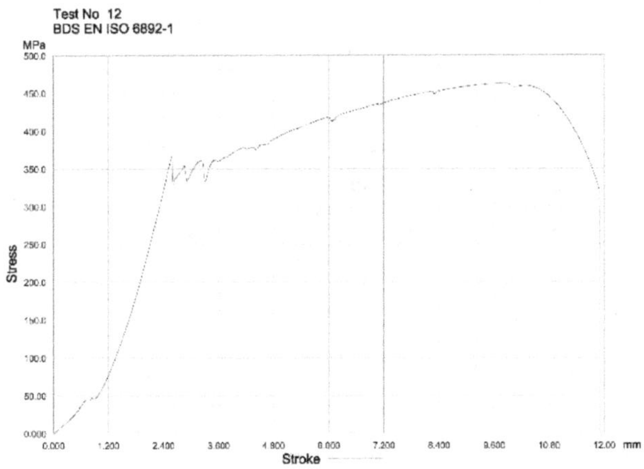

(a)

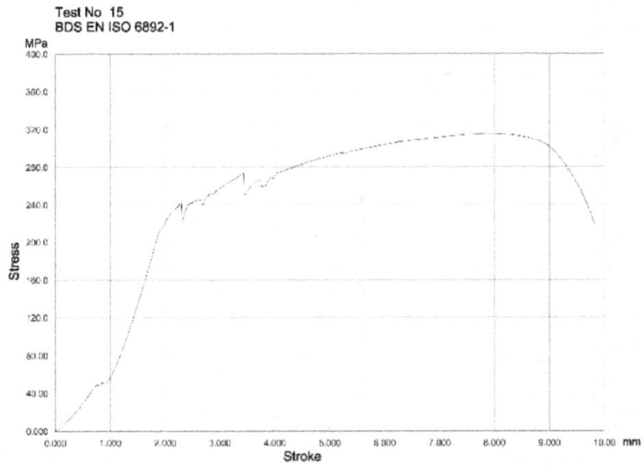

(b)

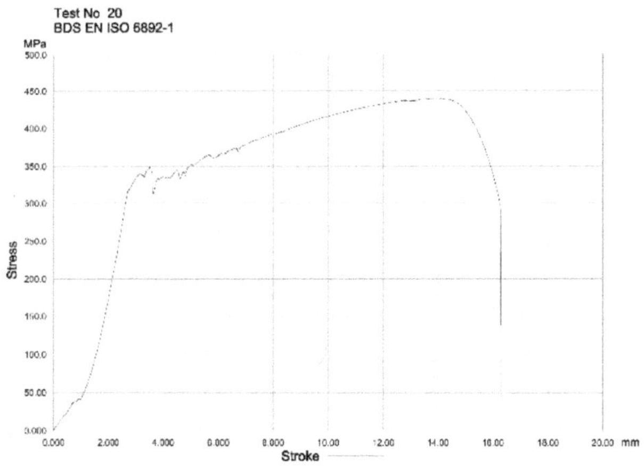

(c)

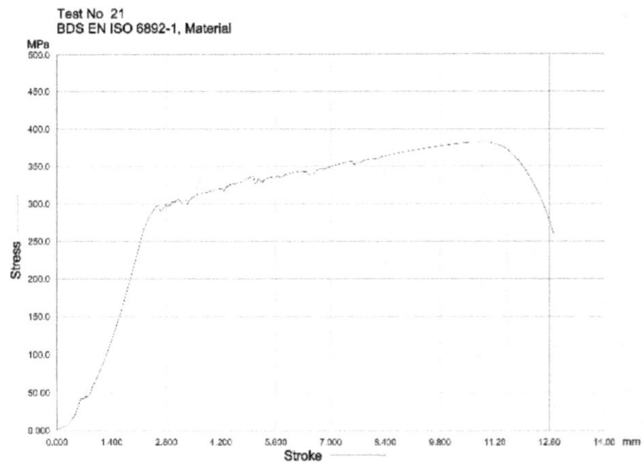

(d)

Figure 5.6.1 Stress-strain curve on corroded steel test specimens; (a) Test № 12 [125]; (b) Test № 15 [125]; (c) Test № 20 [125]; (d) Test № 21 [125];

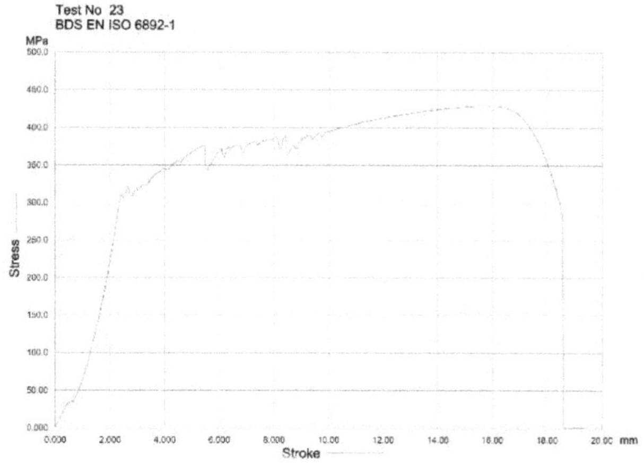

(a)

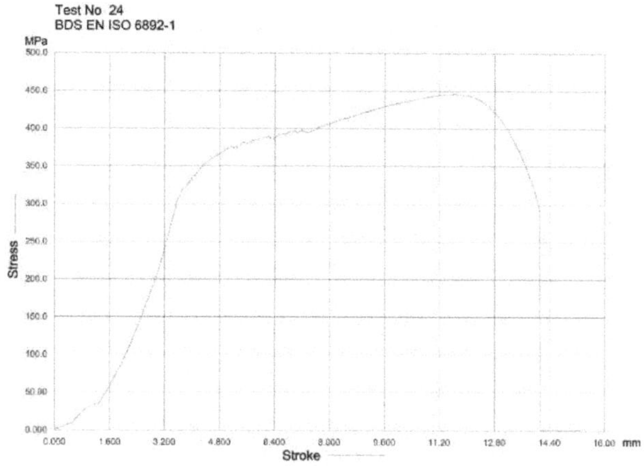

(b)

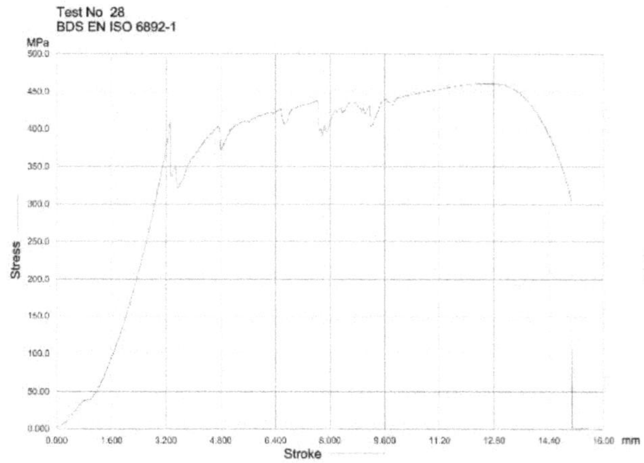

(c)

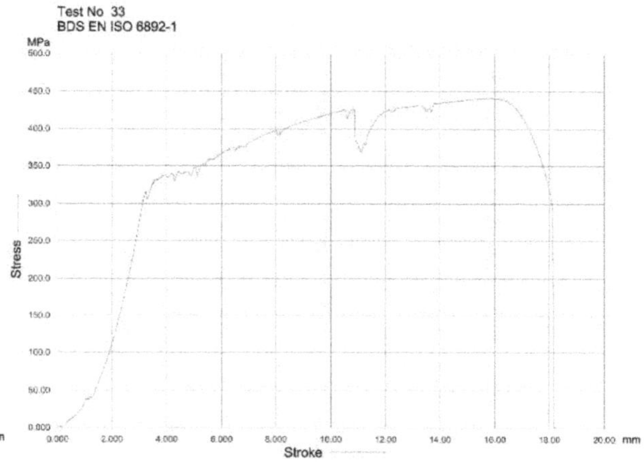

(d)

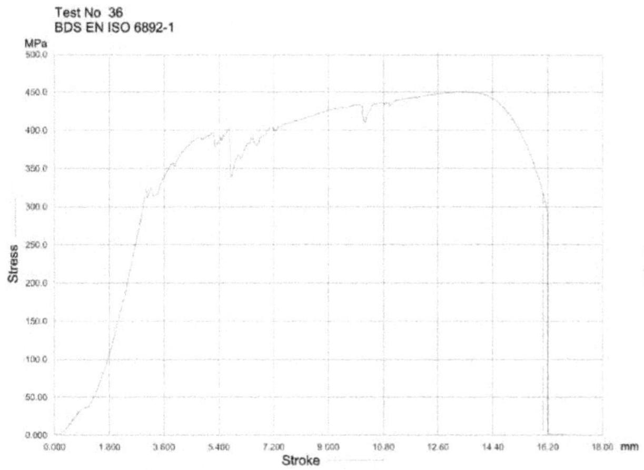

(e)

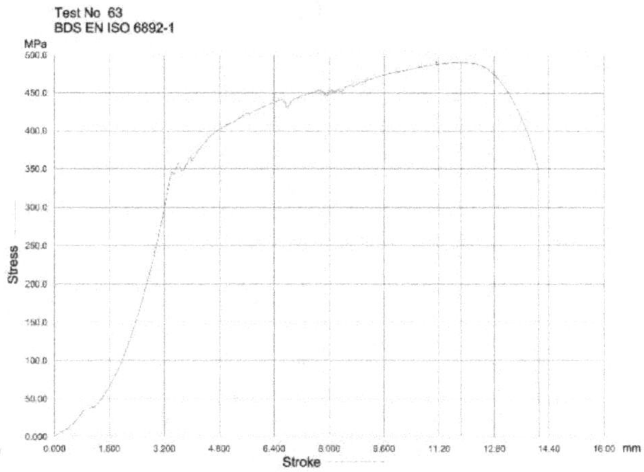

(f)

Figure 5.6.2 Stress-strain curve on corroded steel test specimens; (a) Test № 24 [125]; (b) Test № 23 [125]; (c) Test № 28 [125]; (d) Test № 33 [125]; (e) Test № 36 [125]; (f) Test № 63 [125];

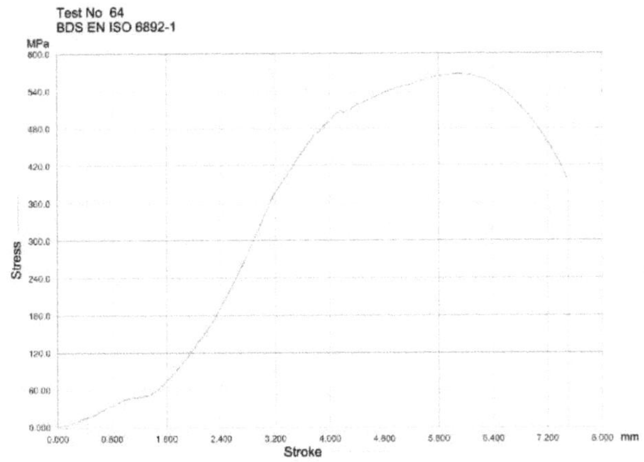

(a)

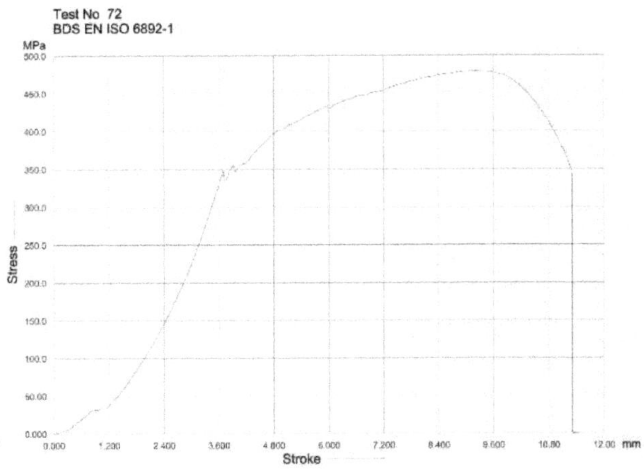

(b)

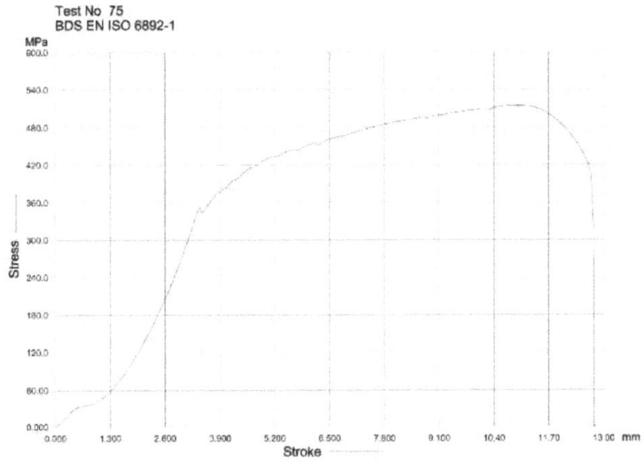

(c)

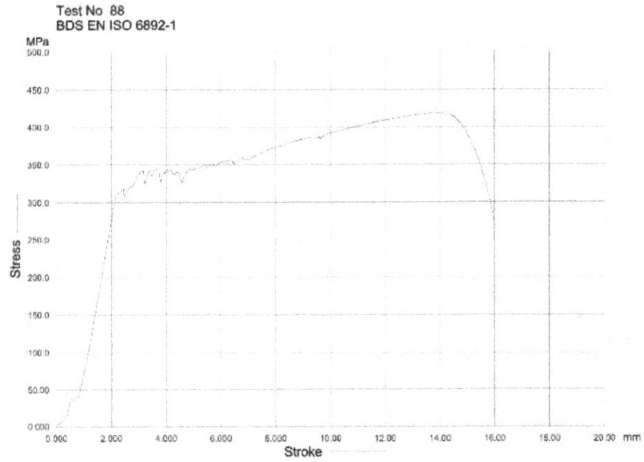

(d)

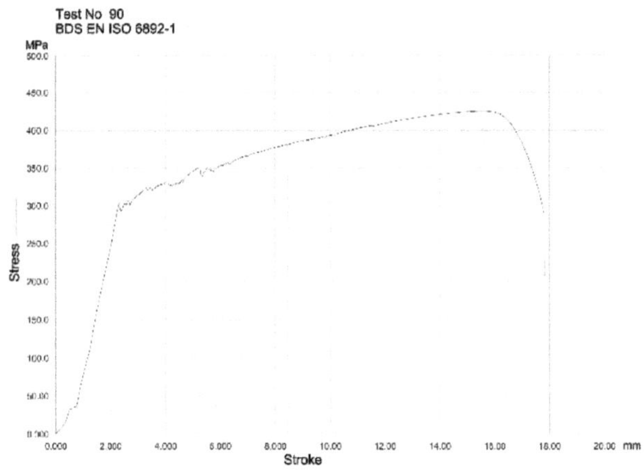

(e)

Figure 5.6.3 Stress-strain curve on corroded steel test specimens; (a) Test № 64 [125]; (b) Test № 72 [125]; (c) Test № 75 [125]; (d) Test № 88 [125]; (e) Test № 90 [125];

Table 5.6.2 Experimental data for change in diameter (in mm) along the length of the element according to the time of corrosion influence [132];

Points	Test specimens №						
	21	12	20	15	90	88	39
1	5.62	5.64	6.18	5.25	6.4	6.56	6.52
2	4.56	5.71	5.54	4.54	5.63	6.10	5.76
4	4.73	5.35	5.59	5.02	5.35	5.71	5.34
5	4.60	5.65	5.78	4.94	5.07	5.94	5.36
6	4.84	5.43	5.62	4.09	4.99	5.96	5.67
7	4.93	4.23	4.64	4.22	4.98	5.72	5.46
8	5.04	4.95	4.59	4.15	4.92	5.39	5.27
9	5.08	4.54	6.09	4.27	5.47	5.20	5.18
10	4.42	5.69	6.14	4.28	5.21	5.24	4.90
11	4.41	5.76	5.82	4.47	6.25	6.11	5.00
12	4.55	5.88	5.62	4.42	6.68	5.28	5.40
13	5.36	5.72	5.81	4.52	5.56	5.68	5.51
14	5.48	5.92	5.73	4.83	5.60	5.34	5.46
15	5.34	6.49	6.01	5.32	5.41	5.57	5.50
16	4.40	5.94	5.78	5.07	5.62	5.56	5.48
17	4.49	6.36	6.36	5.06	5.79	5.30	5.25
18	4.48	6.61	6.05	5.57	5.27	6.00	5.31
19	4.62	6.52	6.76	5.31	5.12	4.97	5.26
20	5.20	6.67	6.95	4.98	5.49	5.41	6.34
21	4.77	6.44	6.59	5.01	6.14	5.09	6.06
22	5.11	6.52	6.73	4.90	6.31	5.06	5.96
23	5.52	5.43	6.51	5.19	6.42	5.09	5.72
24	5.90	5.25	6.62	4.86	6.33	4.67	5.86
25	5.75	5.78	5.71	5.20	6.12	4.85	5.47
26	5.69	5.83	6.03	5.16	6.46	4.94	5.55
27	5.98	5.80	5.80	5.22	6.52	5.25	5.89
28	5.84	5.83	5.48	5.42	6.71	6.57	5.75
29	6.00	5.76	6.07	5.64	7.09	6.38	5.96
30	5.97	6.48	6.36	5.74	6.08	5.18	6.04

Table 5.6.3 Experimental data for change in diameter (in mm) along the length of the element according to the time of corrosion influence [132];

Points	Test specimens №							
	28	33	23	24	75	72	64	63
1	6.42	6.67	6.26	5.95	9.23	6.45	5.94	6.46
2	5.57	5.40	5.55	5.46	7.14	6.28	5.48	5.34
4	5.53	5.12	5.26	5.43	6.98	7.75	5.33	4.98
5	5.12	5.06	5.05	5.27	6.23	6.58	5.14	5.01
6	5.44	4.69	5.51	5.13	6.62	6.15	5.58	6.41
7	5.34	4.58	5.62	5.01	6.98	5.62	5.87	5.97
8	5.42	4.63	5.41	5.08	6.62	5.90	5.95	6.33
9	5.36	4.72	5.51	5.29	6.43	5.83	6.05	5.82
10	5.34	4.80	5.29	5.35	5.44	5.58	6.25	6.02
11	5.63	5.66	5.14	5.50	4.81	5.04	6.57	5.58
12	5.55	4.61	5.13	5.19	5.61	4.95	6.62	5.90
13	5.51	4.72	4.92	4.99	5.72	6.96	6.17	5.64
14	5.35	4.88	5.03	5.30	5.85	6.25	6.42	5.38
15	5.37	4.94	5.34	5.57	6.71	6.39	6.35	5.86
16	5.88	4.87	5.37	5.48	6.95	6.52	4.07	6.26
17	5.49	4.99	5.60	5.28	7.24	6.64	5.83	5.34
18	5.46	5.21	5.21	5.36	6.21	7.34	6.43	5.30
19	5.18	4.59	5.12	5.34	5.27	7.01	6.14	5.28
20	4.93	4.95	4.95	5.35	5.98	6.71	7.26	5.05
21	4.70	5.59	5.21	6.10	6.56	6.53	7.09	4.62
22	4.71	5.87	6.24	5.80	6.12	6.72	7.20	4.80
23	4.78	5.31	5.43	5.93	6.45	6.51	6.27	4.95
24	5.19	5.36	5.47	5.78	6.83	6.86	6.19	5.58
25	5.33	5.58	5.34	5.82	6.78	7.05	5.76	5.70
26	5.71	5.46	5.43	5.89	6.14	6.81	5.62	5.57
27	5.45	5.19	5.60	5.78	7.46	6.85	6.28	5.67
28	5.06	4.94	5.62	5.71	7.44	6.73	6.41	5.89
29	5.11	4.70	5.19	5.35	7.22	7.37	5.73	5.16
30	5.29	4.83	5.26	5.79	8.92	6.24	5.80	6.51

7. Seventh experimental data set

Table 5.7.1 Time of corrosion influence according to corrosion category

Corrosion category C1 [months]	Corrosion category C2 [months]	Corrosion category C3 [months]	Corrosion category C4 [months]	Corrosion category C5 [months]	Corrosion category CX [months]
64615	3360	1680	1050	420	120

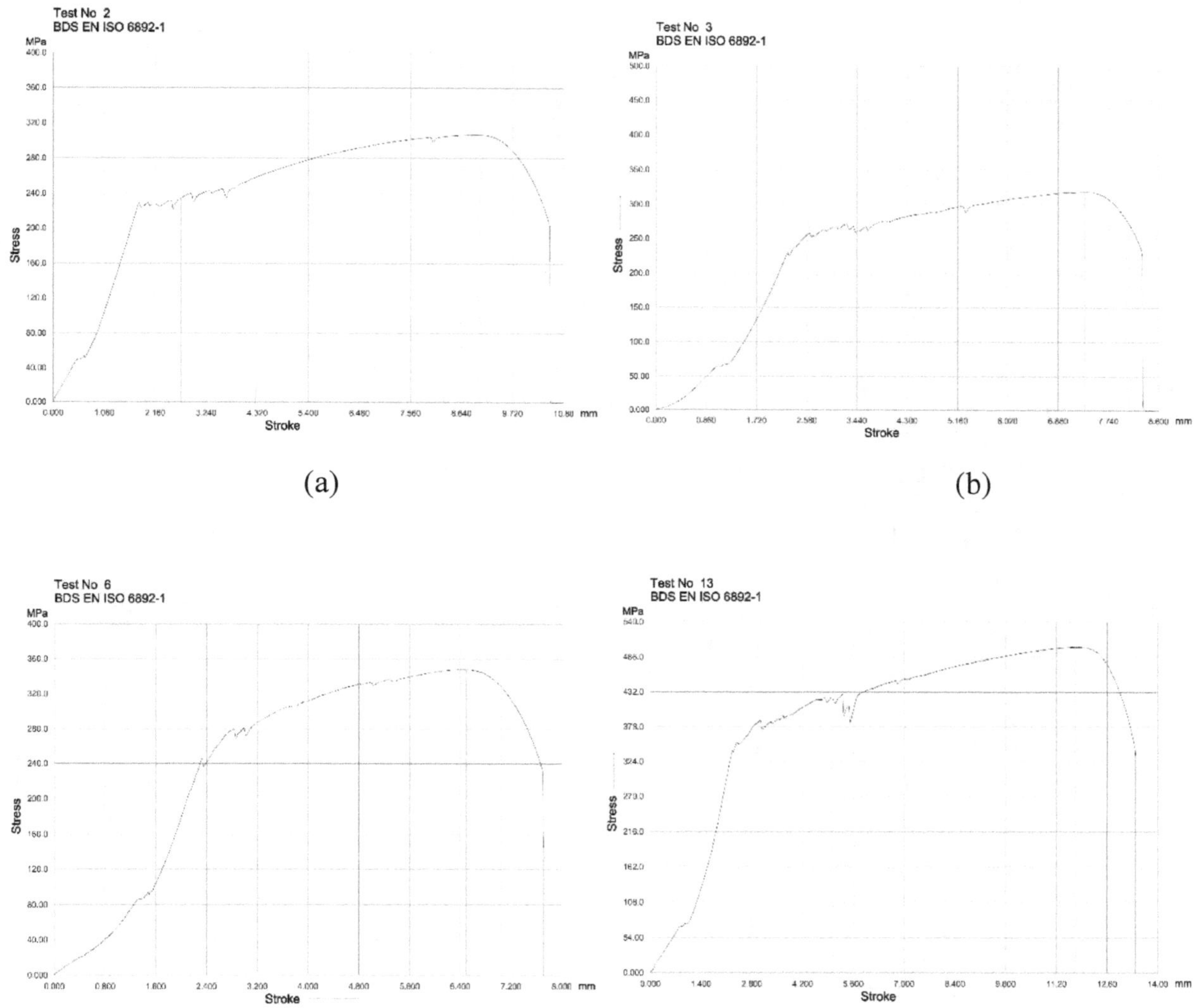

(a) (b)

(c) (d)

Figure 5.7.1 Stress-strain curve on corroded steel test specimens; (a) Test № 2 [126]; (b) Test № 3 [126]; (c) Test № 6 [126]; (d) Test № 13 [126];

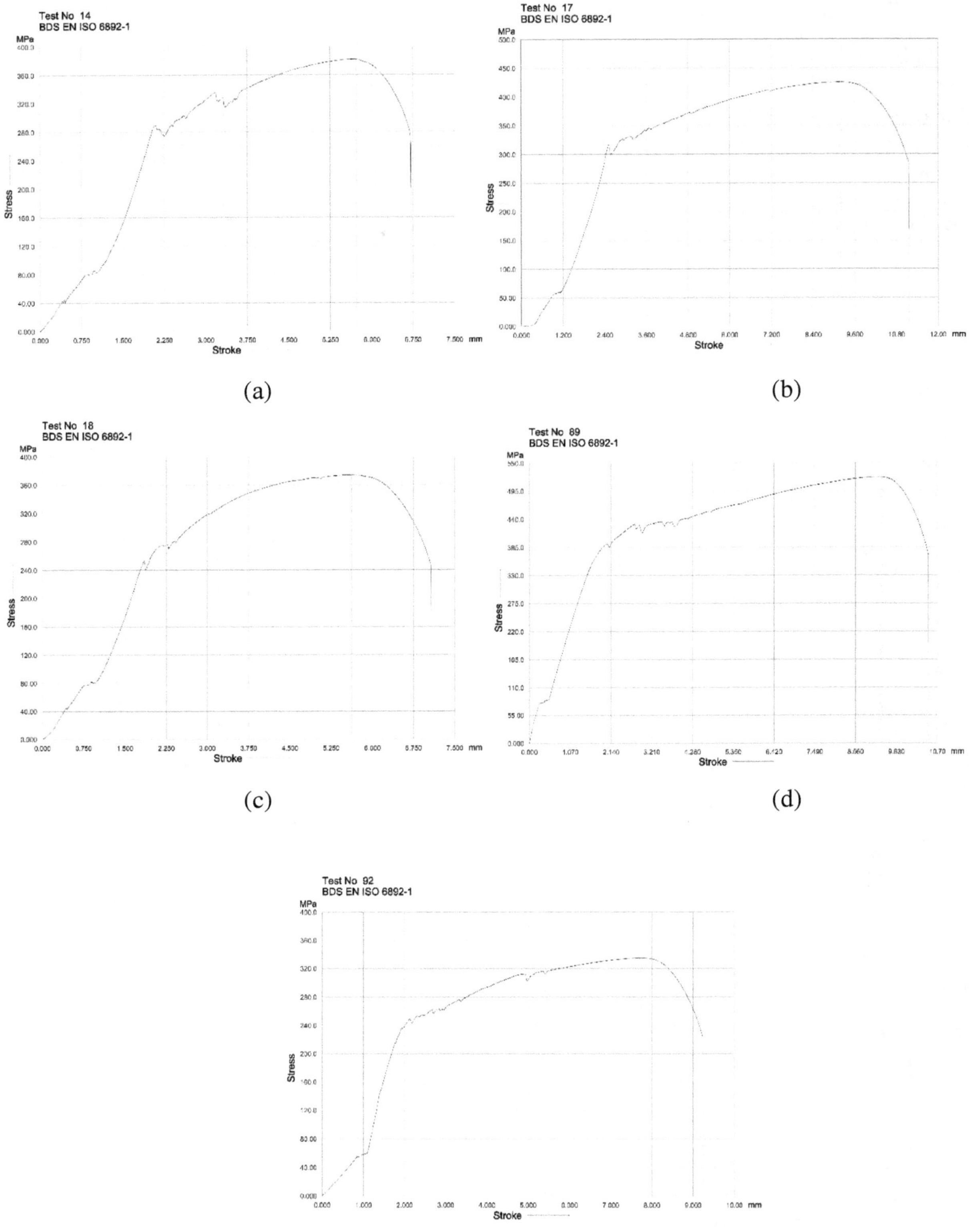

Figure 5.7.2 Stress-strain curve on corroded steel test specimens; (a) Test № 14 [126]; (b) Test № 17 [126]; (c) Test № 18 [126]; (d) Test № 89 [126]; (e) Test № 92 [126];

Table 5.7.2 Experimental data for change in diameter (in mm) along the length of the element according to the time of corrosion influence [133];

Points	Test specimens №							
	17	18	14	13	92	89	6	3
1	5.43	5.74	4.24	5.53	5.59	5.14	4.40	6.20
2	4.24	4.33	3.81	4.30	4.92	3.9	4.09	5.17
4	4.11	4.30	3.35	3.97	4.75	4.52	4.41	4.92
5	3.96	4.11	3.34	4.47	4.52	4.50	4.54	4.57
6	3.98	4.56	3.16	5.08	4.35	4.30	4.20	4.42
7	4.24	4.77	3.68	4.87	4.28	4.57	4.40	4.02
8	3.98	4.20	3.61	4.67	4.42	4.45	4.26	3.91
9	3.84	4.72	3.2	4.69	4.05	5.62	4.19	4.05
10	3.80	4.57	3.65	5.00	3.81	4.56	4.05	3.53
11	5.02	4.09	3.58	4.52	3.94	5.54	3.65	4.17
12	4.48	4.64	3.84	4.68	4.10	5.01	3.09	3.76
13	4.68	4.08	4.69	4.72	4.50	5.17	3.40	4.29
14	5.00	3.23	4.77	4.70	4.76	4.67	3.62	4.19
15	5.06	3.49	4.76	4.34	4.86	4.78	3.44	3.60
16	4.52	3.84	4.94	4.25	5.93	5.51	3.45	3.80
17	5.11	4.20	5.03	4.46	5.76	4.63	4.26	4.06
18	4.34	4.62	4.14	4.11	5.51	4.36	4.49	4.75
19	4.73	4.19	4.15	4.20	5.52	4.59	4.43	4.41
20	5.67	4.38	4.26	3.35	5.54	4.31	4.74	4.32
21	5.94	4.30	4.08	3.49	4.64	4.89	4.26	4.75
22	5.86	4.53	3.79	3.75	4.15	4.58	4.60	4.54
23	4.67	4.88	3.76	4.39	4.61	4.45	4.75	4.45
24	4.79	4.62	3.87	4.74	5.16	4.33	4.84	3.70
25	4.92	4.53	4.42	4.87	5.35	4.62	5.07	3.81
26	4.83	4.60	4.60	5.23	5.19	4.78	4.10	4.38
27	5.23	4.51	4.98	5.50	5.18	5.11	4.03	4.93
28	5.71	4.94	5.05	5.29	4.69	5.25	4.45	4.46
29	5.41	4.97	5.12	5.30	5.23	5.36	6.07	5.09
30	4.81	5.12	4.95	5.45	5.59	6.57	5.78	4.89

8. Eighth experimental data set

Table 5.8.1 Time of corrosion influence according to corrosion category

Corrosion category C1 [months]	Corrosion category C2 [months]	Corrosion category C3 [months]	Corrosion category C4 [months]	Corrosion category C5 [months]	Corrosion category CX [months]
73107	3801	1901	1188	475	136

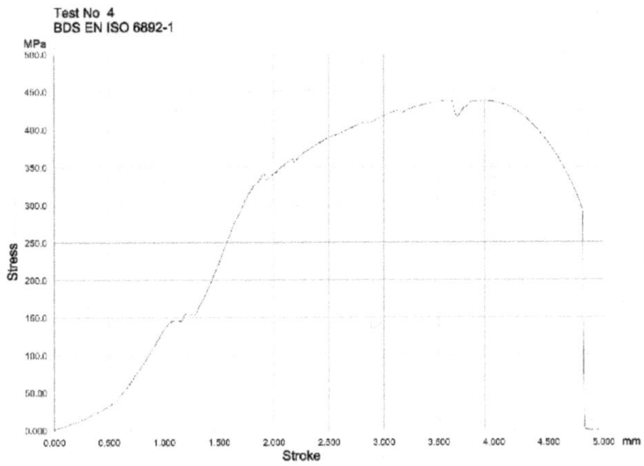

(a)

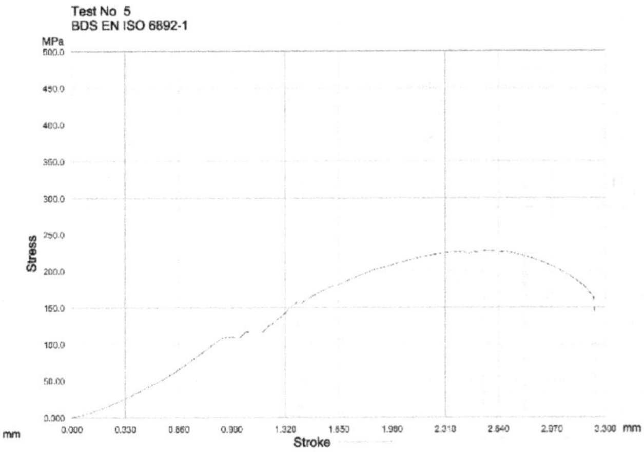

(b)

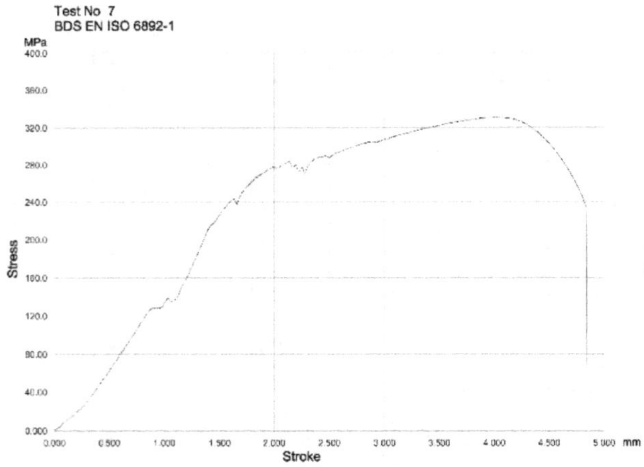

(c)

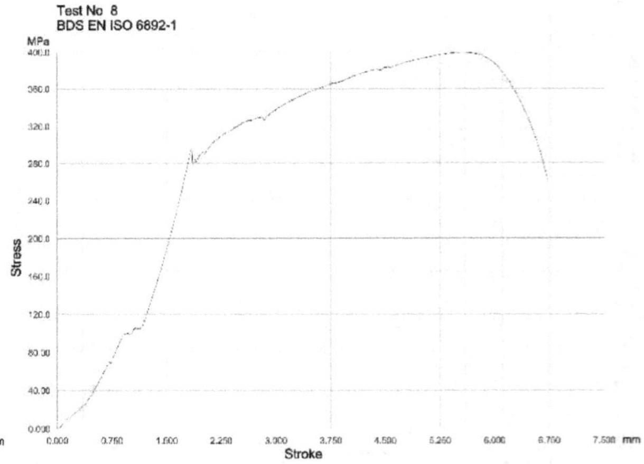

(d)

Figure 5.8.1 Stress-strain curve on corroded steel test specimens; (a) Test № 4 [127]; (b) Test № 5 [127]; (c) Test № 7 [127]; (d) Test № 8 [127];

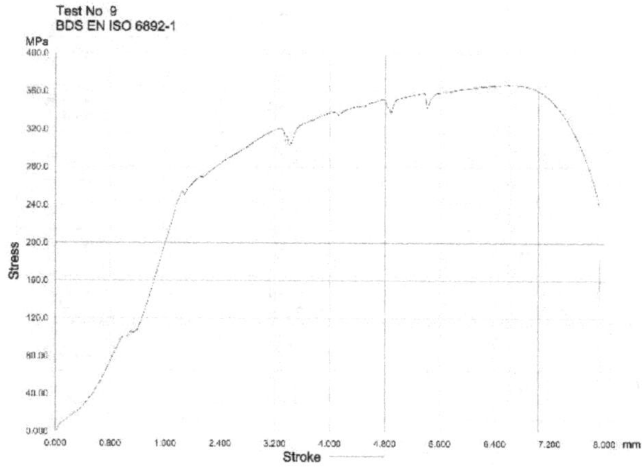

Figure 5.8.2 Stress-strain curve on corroded steel test specimens - Test № 9 [127];

Table 5.8.2 Experimental data for change in diameter (in mm) along the length of the element according to the time of corrosion influence [134];

Points	Test specimens №				
	7	9	5	4	8
1	2.51	3.47	4.15	5.78	3.84
2	2.87	3.10	3.84	4.43	3.27
4	3.38	2.83	3.95	3.84	3.00
5	3.57	4.04	3.59	3.67	3.63
6	3.16	3.84	3.85	4.20	4.31
7	3.11	3.49	4.18	4.18	4.08
8	3.11	3.64	3.84	3.87	3.71
9	3.12	3.47	4.43	3.93	3.06
10	2.61	4.16	4.25	3.81	2.94
11	2.87	3.55	4.04	3.01	2.83
12	4.13	2.95	3.70	3.00	2.82
13	4.84	2.92	2.70	2.33	2.85
14	5.20	3.05	2.80	2.95	4.03
15	4.23	3.17	3.09	3.33	3.37
16	3.67	3.50	3.46	2.83	3.72
17	3.89	3.30	3.23	3.64	3.76
18	4.13	3.50	4.12	3.66	3.64
19	4.13	3.19	4.44	4.12	3.80
20	4.47	4.36	4.08	4.23	4.36

21	3.86	4.33	3.06	4.35	4.07
22	4.16	4.69	3.74	3.52	4.18
23	4.53	4.06	3.36	3.17	4.62
24	5.22	4.99	4.17	3.53	4.16
25	5.53	4.92	3.73	3.51	3.73
26	5.71	4.33	3.68	3.63	3.30
27	5.55	3.73	3.61	4.25	3.53
28	5.05	4.13	2.96	3.99	4.13
29	4.94	4.10	3.35	3.71	4.53
30	5.13	4.81	3.45	4.08	4.44

9. Ninth experimental data set

Table 5.9.1 Time of corrosion influence according to corrosion category

Corrosion category C1 [months]	Corrosion category C2 [months]	Corrosion category C3 [months]	Corrosion category C4 [months]	Corrosion category C5 [months]	Corrosion category CX [months]
77262	4018	2009	1259	502	143

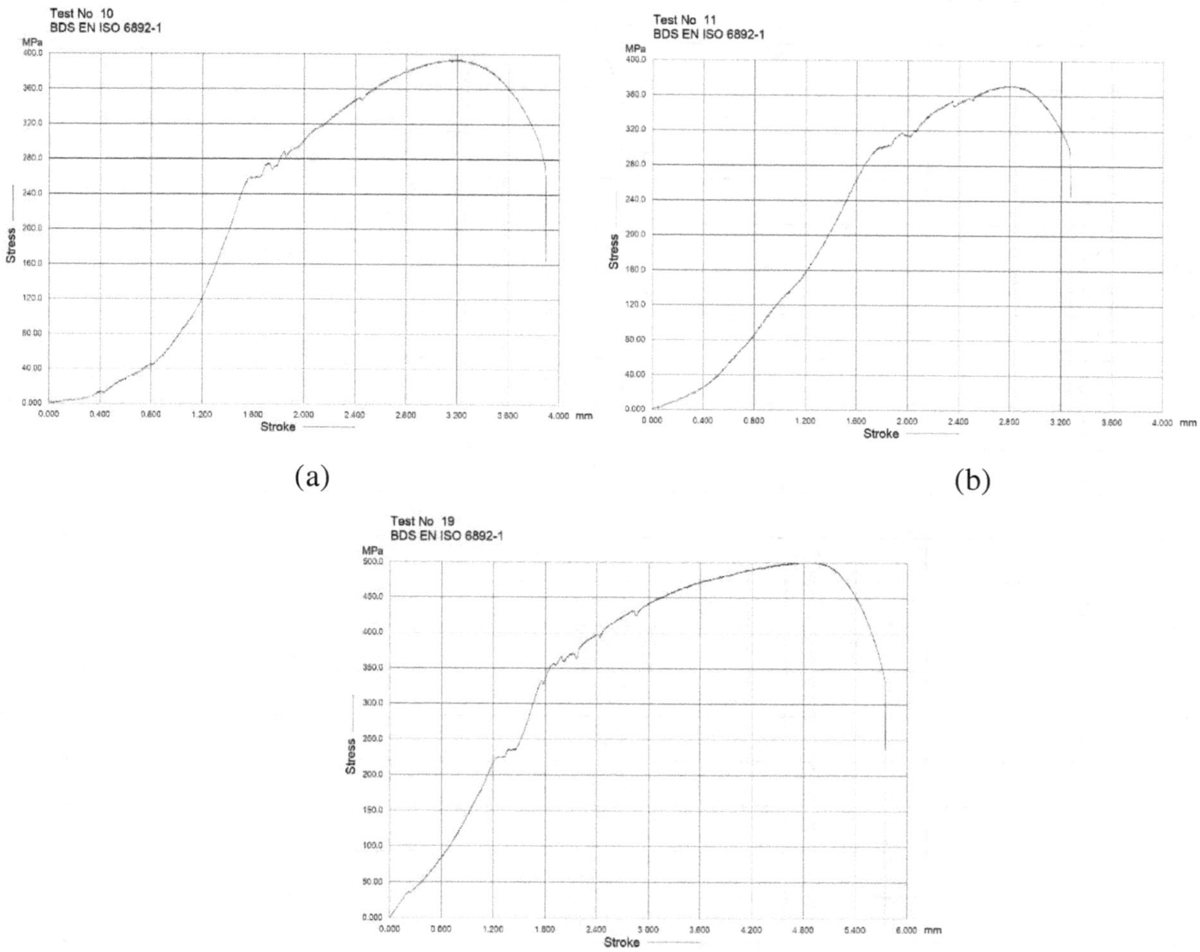

(a) (b) (c)

Figure 5.9.1 Stress-strain curve on corroded steel test specimens; (a) Test № 10 [128]; (b) Test № 11 [128]; (c) Test № 19 [128];

Table 5.9.2 Experimental data for change in diameter (in mm) along the length of the element according to the time of corrosion influence [135];

Points	Test specimens №		
	19	10	11
1	4.92	4.95	4.63
2	3.90	4.20	3.42
4	3.98	4.22	3.42
5	3.67	4.69	2.79
6	3.87	3.24	2.37
7	3.52	3.22	2.23
8	4.08	3.88	2.33
9	3.56	3.44	2.38
10	3.27	3.19	3.28
11	2.65	3.26	3.44
12	2.73	2.07	3.71
13	1.91	1.77	3.08
14	2.66	2.32	3.43
15	2.06	3.20	2.29
16	2.76	3.02	1.63
17	2.76	3.18	2.53
18	3.45	3.78	2.94
19	4.13	4.58	3.65
20	4.11	3.67	3.31
21	3.34	3.67	2.89
22	3.42	3.34	2.66
23	3.40	2.86	2.77
24	2.72	2.93	2.62
25	2.72	3.75	2.81
26	3.05	4.76	3.03
27	3.48	4.45	3.23
28	4.88	4.57	4.00
29	4.09	3.71	3.73
30	4.90	4.55	3.56

REFERENCES

[1] AGGARWAL, Jagdishkumar Keshoram. Notes on nonlinear systems. VNR Co., 1972.

[2] AREF, Hassan. Stochastic particle motion in laminar flows. Physics of Fluids A: Fluid Dynamics, 1991, 3.5: 1009-1016.

[3] CAMBEL, Ali Bulent. Applied chaos theory: A paradigm for complexity. Elsevier, 1993.

[4] FEIGENBAUM, Mitchell J. Universal behavior in nonlinear systems. Physica D: Nonlinear Phenomena, 1983, 7.1-3: 16-39.

[5] GRASSBERGER, Peter; SCHREIBER, Thomas; SCHAFFRATH, Carsten. Nonlinear time sequence analysis. International Journal of Bifurcation and Chaos, 1991, 1.03: 521-547.

[6] LORENZ, Edward N. Atmospheric predictability experiments with a large numerical model. Tellus, 1982, 34.6: 505-513.

[7] LORENZ, Edward N. Predictability: A problem partly solved. In: Proc. Seminar on predictability. 1996. p. 1-18.

[8] SHEN, Bo-Wen. A Review of Lorenz's Models from 1960 to 2008. International Journal of Bifurcation and Chaos, 2023, 33.10: 2330024.

[9] RICHTMYER, Robert D.; BURDORF, Christopher. Principles of advanced mathematical physics. New York: Springer, 1978.

[10] SUGIHARA, George; MAY, Robert M. Nonlinear forecasting as a way of distinguishing chaos from measurement error in time series. Nature, 1990, 344.6268: 734-741.

[11] YAMAGUTI, Masaya; USHIKI, Shigehiro. Chaos in numerical analysis of ordinary differential equations. Physica D: Nonlinear Phenomena, 1981, 3.3: 618-626.

[12] ABRAHAM, Bovas; LEDOLTER, Johannes. Statistical methods for forecasting. John Wiley & Sons, 2009.

[13] POINCARÉ, Henri. Leçons sur la théorie mathématique de la lumière: professées pendant le premier semestre 1887-1888. G. Carré, 1889. [in French]

[14] ЛОРАН, П.-Ж. Аппроксимация и оптимизация. Мир, Москва, 1975. [in Russian]

[15] ВАЗАН, М. Т.; ВАЙСВОРД, Э. М. Стохастическая аппроксимация. Мир, Москва 1972. [in Russian]

[16] ADVANI, Madhu; LAHIRI, Subhaneil; GANGULI, Surya. Statistical mechanics of complex neural systems and high dimensional data. Journal of Statistical Mechanics: Theory and Experiment, 2013, 2013.03: P03014.

[17] HUANG, Haiping. Statistical mechanics of neural networks. Singapore: Springer, 2021.

[18] JAYNES, Edwin T. Information theory and statistical mechanics. Physical review, 1957, 106.4: 620.

[19] HUANG, Kerson. Introduction to statistical physics. Chapman and Hall/CRC, 2009.

[20] EBELING, Werner; SOKOLOV, Igor. Statistical thermodynamics and stochastic theory of nonequilibrium systems. World Scientific Publishing Company, 2005.

[21] GIBBS, Josiah Willard. On the fundamental formula of statistical mechanics, with applications to astronomy and thermodynamics. In: Proceedings of the American Association for the Advancement of Science. 1884. p. 57-58.

[22] GIBBS, Josiah Willard. Elementary principles in statistical mechanics: developed with especial reference to the rational foundations of thermodynamics. C. Scribner's sons, 1902.

[23] TOLMAN, Richard Chace. The principles of statistical mechanics. Courier Corporation, 1979.

[24] GAO, Xiang. The mathematics of the ensemble theory. Results in Physics, 2022, 34: 105230.

[25] GORBAN, Alexander N.; TYUKIN, Ivan Yu. Blessing of dimensionality: mathematical foundations of the statistical physics of data. Philosophical Transactions of the Royal Society A: Mathematical, Physical and Engineering Sciences, 2018, 376.2118: 20170237.

[26] BAXTER, Rodney J. Exactly solved models in statistical mechanics. Elsevier, 2016.

[27] BALESCU, Radu. Equilibrium and nonequilibrium statistical mechanics. NASA STI/Recon Technical Report A, 1975, 76: 32809.

[28] GRANDY JR, Walter T. Foundations of statistical mechanics: Equilibrium theory. Springer Science & Business Media, 2012.

[29] HEER, Clifford V. Statistical mechanics, kinetic theory, and stochastic processes. Elsevier, 2012.

[30] BRUSH, Stephen G. Foundations of Statistical Mechanics 1845—1915. Archive for History of Exact Sciences, 1967, 4.3: 145-183.

[31] COWAN, Brian. Topics in statistical mechanics. World Scientific, 2021.

[32] HILL, Terrell L. Statistical mechanics: principles and selected applications. Courier Corporation, 2013.

[33] ALEKSANDR, I.; KHINCHIN, Akovlevich. Mathematical foundations of statistical mechanics. Courier Corporation, 1949.

[34] WEINER, Jerome Harris. Statistical mechanics of elasticity. Courier Corporation, 2012.

[35] FRIGG, Roman; WERNDL, Charlotte. Philosophy of Statistical Mechanics. 2023.

[36] HOBSON, Art. Concepts in statistical mechanics. CRC Press, 1987.

[37] CHANDLER, David. Introduction to modern statistical. Mechanics. Oxford University Press, Oxford, UK, 1987, 5.449: 11.

[38] JAYNES, Edwin T. Foundations of probability theory and statistical mechanics. In: Delaware seminar in the foundations of physics. Berlin, Heidelberg: Springer Berlin Heidelberg, 1967. p. 77-101.

[39] PRIGOGINE, Ilya. Non-equilibrium statistical mechanics. Courier Dover Publications, 2017.

[40] NAVARRO, Luis. Gibbs, Einstein and the foundations of statistical mechanics. Archive for history of exact sciences, 1998, 53: 147-180.

[41] BRUSH, Stephen G. Statistical mechanics and the philosophy of science: some historical notes. In: PSA: Proceedings of the Biennial Meeting of the Philosophy of Science Association. Cambridge University Press, 1976. p. 551-584.

[42] RYABOV, Valeriy A. Principles of Statistical Physics and Numerical Modeling. IOP Publishing, 2018.

[43] LIKHAREV, Konstantin K. Statistical Mechanics: Lecture notes: Lecture notes. IOP Publishing, 2019.

[44] COHEN, E. G. D. Boltzmann and statistical mechanics. In: Dynamics: Models and Kinetic Methods for Non-equilibrium Many Body Systems. Dordrecht: Springer Netherlands, 2000. p. 221-238.

[45] TOLMAN, Richard Chace. The principles of statistical mechanics. Courier Corporation, 1979.

[46] JAYNES, E. T. Predictive statistical mechanics. In: Frontiers of Nonequilibrium Statistical Physics. Boston, MA: Springer US, 1986. p. 33-55.

[47] FRIGG, Roman; WERNDL, Charlotte. Statistical mechanics: A tale of two theories. The Monist, 2019, 102.4: 424-438.

[48] BAGCHI, Biman. Statistical mechanics for chemistry and materials science. CRC Press, 2018.

[49] RYSSELBERGHE, Pierre Van. Mathematical foundations of statistical mechanics. 1949.

[50] BAGCHI, Biman. Nonequilibrium Statistical Mechanics: An Introduction with Applications. Chapman and Hall/CRC, 2024.

[51] CERCIGNANI, Carlo; CERCIGNANI, Carlo. Basic Principles. Mathematical Methods in Kinetic Theory, 1990, 1-39.

[52] ELLIS, Richard S. An overview of the theory of large deviations and applications to statistical mechanics. Scandinavian Actuarial Journal, 1995, 1995.1: 97-142.

[53] FREUDENTHAL, Alfred M. Safety and the probability of structural failure. Transactions of the American Society of Civil Engineers, 1956, 121.1: 1337-1375.

[54] JOHNSON, Arne I. Strength, safety and economical dimensions of structures. 1953.

[55] SHINOZUKA, Masanobu. Methods of safety and reliability analysis. In: International Conference on Structural Safety and Reliability. Pergamon, 1972. p. 11-45.

[56] MAYER, Max. Die Sicherheit der Bauwerke und ihre Berechnung nach Grenzkräften anstatt nach zulässigen Spannungen. J. Springer, 1926. [in German]

[57] BOLOTIN, Vladimir V. Mechanics of fatigue. Crc Press, 2020.

[58] BOLOTIN, V. V. Stochastic boundary-value problems in the theory of plates and shells. In: Proc. Sixth All-Union Conf. Theory of Plates and Shells. Nauka, Moscow. 1966 [in Russian].

[59] РЖАНИЦЫН, А.Р. Расчет сооружений с учетом пластических свойств материалов. Стройиздат, Москва, 1954 [in Russian]

[60] STRELETSKY, N. S. Statistical basic of the safety factor of structures, Stroyizdat, Moscow, 1947 [in Russian]

[61] ХОЦИАЛОВ, Н. Ф. Запасы прочности. Строительная промышленность, 1929, 10: 840. [in Russian]

[62] THOFI-CHRISENSEN, P. and Baker, M. J. Structural reliability theory and its applications, Springer Verlag, Berlin, 1982

[63] PAIKOWSKY, Samuel G. Load and resistance factor design (LRFD) for deep foundations. Transportation Research Board, 2004.

[64] REISS, Rolf-Dieter; THOMAS, Michael; REISS, R. D. Statistical analysis of extreme values. Basel: Birkhäuser, 1997.

[65] БОЛОТИН, В. Методы теории вероятностей и теории надежности в рачетах сооружений, Стройиздат, Москва, 1981 [in Russian]

[66] КИСЛЯКОВ, С. Стохастически методи в приложната механика, Техника, София, 1981 [in Bulgarian]

[67] GRIMMETT, Geoffrey; STIRZAKER, David. Probability and random processes. Oxford university press, 2020.

[68] KORALOV, Leonid; SINAI, Yakov G. Theory of probability and random processes. Springer Science & Business Media, 2007.

[69] IBE, Oliver. Fundamentals of applied probability and random processes. Academic Press, 2014.

[70] ROZANOV, Yu. Probability theory, random processes and mathematical statistics. Springer Science & Business Media, 2012.

[71] GRAMI, Ali. Probability, random variables, statistics, and random processes: Fundamentals & applications. John Wiley & Sons, 2019.

[72] MAISEL, Louis. Probability, statistics and random processes. Simon and Schuster, 1971.

[73] MILLER, Irwin. Probability, random variables, and stochastic processes. 1966.

[74] PROAKIS, J. Probability, random variables and stochastic processes. IEEE Transactions on Acoustics, Speech, and Signal Processing, 1985, 33.6: 1637-1637.

[75] GHAHRAMANI, Saeed. Fundamentals of probability: with stochastic processes. Chapman and Hall/CRC, 2018.

[76] SNYDER, Donald L.; MILLER, Michael I. Random point processes in time and space. Springer Science & Business Media, 2012.

[77] TSE, Yiu-Kuen. Nonlife actuarial models: theory, methods and evaluation. Cambridge University Press, 2023.

[78] FOX, John. Linear mixed models. Appendix to an R and S-plus Companion to Applied Regression, 2002, 16: 2349-2380.

[79] STROUP, Walter W. Generalized linear mixed models: modern concepts, methods and applications. CRC press, 2012.

[80] BOLKER, Benjamin M. Linear and generalized linear mixed models. Ecological statistics: contemporary theory and application, 2015, 2015: 309-333.

[81] WELHAM, Sue, et al. Prediction in linear mixed models. Australian & New Zealand Journal of Statistics, 2004, 46.3: 325-347.

[82] BRESLOW, Norman E.; CLAYTON, David G. Approximate inference in generalized linear mixed models. Journal of the American statistical Association, 1993, 88.421: 9-25.

[83] CHIHARA, Laura M.; HESTERBERG, Tim C. Mathematical statistics with resampling and R. John Wiley & Sons, 2018.

[84] YU, Chong Ho. Resampling methods: concepts, applications, and justification. Practical Assessment, Research, and Evaluation, 2019, 8.1: 19.

[85] JAMES, Gareth, et al. Resampling methods. In: An introduction to statistical learning: With applications in python. Cham: Springer International Publishing, 2023. p. 201-228.

[86] VAN DER WAERDEN, Bartel Leendert. Mathematical statistics. Springer Science & Business Media, 2013.

[87] LARSEN, Richard J.; MARX, Morris L. An introduction to mathematical statistics. Hoboken, NJ: Prentice Hall, 2005.

[88] GUPTA, S. C.; KAPOOR, V. K. Fundamentals of mathematical statistics. Sultan Chand & Sons, 2020.

[89] THOMPSON, Colin J. Mathematical statistical mechanics. Princeton University Press, 2015.

[90] ZAĬTSEV, G. N. The mathematical statistics in experimental botany. 1984.

[91] SERFLING, Robert J. Approximation theorems of mathematical statistics. John Wiley & Sons, 2009.

[92] TURCHIN, Valentin Fedorovich; KOZLOV, Viktor Pavlovich; MALKEVICH, Matvei Semenovich. The use of mathematical-statistics methods in the solution of incorrectly posed problems. Soviet Physics Uspekhi, 1971, 13.6: 681.

[93] PUGACHEV, Vladimir Semenovich. Probability theory and mathematical statistics for engineers. Elsevier, 2014.

[94] MANOUKIAN, Edward B. Modern concepts and theorems of mathematical statistics. Springer Science & Business Media, 2012.

[95] RUSKEEPAA, Heikki. Mathematica navigator: mathematics, statistics and graphics. Academic Press, 2009.

[96] GELBAUM, Bernard R., et al. Problems in probability theory, mathematical statistics and theory of random functions. Courier Corporation, 1978.

[97] ALEXANDER, Howard W. Elements of mathematical statistics. 1961.

[98] MANOUKIAN, Edward B. Mathematical nonparametric statistics. Taylor & francis, 2022.

[99] BILLINGSLEY, Patrick. Statistical methods in Markov chains. The annals of mathematical statistics, 1961, 12-40.

[100] COVER, Thomas M. Hypothesis testing with finite statistics. The Annals of Mathematical Statistics, 1969, 40.3: 828-835.

[101] BERLINER, L. Mark. Statistics, probability and chaos. Statistical Science, 1992, 69-90.

[102] ЧАЛЪКОВ, В. „Математическо моделиране на сложни системи“, Академия на МВР, София, 2004 [in Bulgarian]

[103] TONG, Howell. Threshold models in non-linear time series analysis. Springer Science & Business Media, 2012.

[104] OZAKI, Tohru; THOMSON, Peter. A non-linear dynamic model for multiplicative seasonal-trend decomposition. Journal of Forecasting, 2002, 21.2: 107-124.

[105] NEUSSER, Klaus. An investigation into a non-linear stochastic trend model. Empirical Economics, 1999, 24: 135-153.

[106] VOGEL, Kristin, et al. Natural hazards in a changing world: Methods for analyzing trends and non-linear changes. Earth's Future, 2024, 12.5: e2023EF003553.

[107] SAHOO, G. B.; SCHLADOW, S. G.; REUTER, J. E. Forecasting stream water temperature using regression analysis, artificial neural network, and chaotic non-linear dynamic models. Journal of hydrology, 2009, 378.3-4: 325-342.

[108] ROPKINS, Karl, et al. Application of non-linear time-alignment and integration methods to environmental time series. TrAC Trends in Analytical Chemistry, 2009, 28.3: 373-391.

[109] STOJKOVIĆ, Milan, et al. Multi-temporal analysis of mean annual and seasonal stream flow trends, including periodicity and multiple non-linear regression. Water resources management, 2014, 28: 4319-4335.

[110] TSAY, Ruey S.; CHEN, Rong. Nonlinear time series analysis. John Wiley & Sons, 2018.

[111] CHEN, Li; GAO, Jiti; VAHID, Farshid. Nonlinear Trending Time Series: Theory and Practice. 2025.

[112] DAWSON, T., et al. EN380 Naval Materials Science and Engineering Course Notes. US Naval Academy, 2007, 1-250.

[113] КИСЛЯКОВ, С. „Съпротивление на материалите“, Техника, София, 1976 [in Bulgarian]

[114] SHOPOV, A. Computation on Corrosion Influence in Ultimate Strength and Strain Depending on Time. International Journal of Mechanical and Mechatronics Engineering, 19.3, 2019, pp. 112-120

[115] SHOPOV, A. Calculation on Yield Strain Depending on Time of Corrosion Influence. International Journal of Innovative Technology and Exploring Engineering, 8.7, 2019, pp. 2391-2396

[116] SHOPOV, A. Theoretical-Computation Conception for Forecasting on Corrosion Influence into Steel Elements at Sustainable Development. International Journal of Innovative Technology and Exploring Engineering, 8.8, 2019, pp. 2253-2261

[117] SHOPOV, A. and Ganev, Sv. “Forecasting the change of yield stress point on steel with corrosion according to corrosion influence for military shelters, In: 9-th International Scientific Conference on Defensive Technologies "OTEH 2020", Belgrad, Serbia, 2020, pp. 469-474.

[118] SHOPOV, A. Establishing equations for calculating the change of lower yield point depending on the time of corrosion effect. Journal of Mechanics of Continua and Mathematical Sciences, 19.7, 2024, pp.43-58

[119] SHOPOV, A. Equations for calculating and forecasting the change of the proportional limit based on the time of corrosion effect. In: International Conference on Manufacturing, Material science and Engineering “ICMMSE-2024”, Hyderabad, India, 2024 (in press).

[120] SHOPOV, A. Experimental stress-strain curves on steel S355JR specimens with corrosion - after 9139 months corrosion influence. DOI:10.6084/m9.figshare.28700573

[121] SHOPOV, A. Experimental stress-strain curves on steel S355JR specimens with corrosion - after 14769 months corrosion influence. DOI:10.6084/m9.figshare.28700624

[122] SHOPOV, A. Experimental stress-strain curves on steel S355JR specimens with corrosion - after 24923 months corrosion influence. DOI:10.6084/m9.figshare.28700642

[123] SHOPOV, A. Experimental stress-strain curves on steel S355JR specimens with corrosion - after 34892 months corrosion influence. DOI:10.6084/m9.figshare.28700651

[124] SHOPOV, A. Experimental stress-strain curves on steel S355JR specimens with corrosion - after 46154 months corrosion influence. DOI:10.6084/m9.figshare.28700660

[125] SHOPOV, A. Experimental stress-strain curves on steel S355JR specimens with corrosion - after 54000 months corrosion influence. DOI:10.6084/m9.figshare.28700675

[126] SHOPOV, A. Experimental stress-strain curves on steel S355JR specimens with corrosion - after 64615 months corrosion influence. DOI:10.6084/m9.figshare.28700690

[127] SHOPOV, A. Experimental stress-strain curves on steel S355JR specimens with corrosion - after 73107 months corrosion influence. DOI:10.6084/m9.figshare.28700723

[128] SHOPOV, A. Experimental stress-strain curves on steel S355JR specimens with corrosion - after 77262 months corrosion influence. DOI:10.6084/m9.figshare.28700741

[129] SHOPOV, A. Experimental data of changes on diameter on steel S355JR specimens with corrosion - after 24923 months corrosion influence. DOI:10.6084/m9.figshare.29166851

[130] SHOPOV, A. Experimental data of changes on diameter on steel S355JR specimens with corrosion - after 34892 months corrosion influence. DOI:10.6084/m9.figshare.29166857

[131] SHOPOV, A. Experimental data of changes on diameter on steel S355JR specimens with corrosion - after 46154 months corrosion influence. DOI:10.6084/m9.figshare.29166881

[132] SHOPOV, A. Experimental data of changes on diameter on steel S355JR specimens with corrosion - after 54000 months corrosion influence. DOI:10.6084/m9.figshare.29166896

[133] SHOPOV, A. Experimental data of changes on diameter on steel S355JR specimens with corrosion - after 64615 months corrosion influence. DOI:10.6084/m9.figshare.29166902

[134] SHOPOV, A. Experimental data of changes on diameter on steel S355JR specimens with corrosion - after 73107 months corrosion influence. DOI:10.6084/m9.figshare.29166908

[135] SHOPOV, A. Experimental data of changes on diameter on steel S355JR specimens with corrosion - after 77262 months corrosion influence. DOI:10.6084/m9.figshare.29166911

www.ingramcontent.com/pod-product-compliance
Lightning Source LLC
LaVergne TN
LVHW082339200926
848481LV00012B/3389
9798230302766